The Hardware Startup

Renee DiResta, Brady Forrest, and Ryan Vinyard

Beijing · Cambridge · Farnham · Köln · Sebastopol · Tokyo

The Hardware Startup

by Renee DiResta, Brady Forrest, and Ryan Vinyard

Copyright © 2015 Renee DiResta, Brady Forrest, and Ryan Vinyard. All rights reserved.

Printed in the United States of America.

Published by O'Reilly Media, Inc., 1005 Gravenstein Highway North, Sebastopol, CA 95472.

O'Reilly books may be purchased for educational, business, or sales promotional use. Online editions are also available for most titles (*http://safaribooksonline.com*). For more information, contact our corporate/institutional sales department: 800-998-9938 or *corporate@oreilly.com*.

Editors: Brian Sawyer and Mike Loukides
Production Editor: Matthew Hacker
Copyeditor: Molly Ives Brower
Proofreader: Eileen Cohen
Indexer: WordCo Indexing Services, Inc.
Interior Designer: David Futato
Cover Designer: Edie Freedman
Illustrator: Renee DiResta, Brady Forrest, and Ryan Vinyard

June 2015: First Edition

Revision History for the First Edition

2015-05-18: First Release

See *http://oreilly.com/catalog/errata.csp?isbn=9781449371036* for release details.

The O'Reilly logo is a registered trademark of O'Reilly Media, Inc. *The Hardware Startup*, the cover image, and related trade dress are trademarks of O'Reilly Media, Inc.

While the publisher and the authors have used good faith efforts to ensure that the information and instructions contained in this work are accurate, the publisher and the authors disclaim all responsibility for errors or omissions, including without limitation responsibility for damages resulting from the use of or reliance on this work. Use of the information and instructions contained in this work is at your own risk. If any code samples or other technology this work contains or describes is subject to open source licenses or the intellectual property rights of others, it is your responsibility to ensure that your use thereof complies with such licenses and/or rights.

978-1-449-37103-6

[LSI]

Table of Contents

Preface. ix

1. **The Hardware Startup Landscape**. 1
   Early Makers   1
      The Whole Earth Catalog   2
      Communities Around New Technology   2
      MIT Center for Bits and Atoms   3
      Make Magazine   3
   Technology Enables Scale   4
      Rapid Prototyping   5
      Inexpensive Components   5
      Small-Batch Manufacturing   6
      Open Source Hardware   6
      Online Community   7
      The Supplemental Ecosystem   7
      The "Lean Startup" and Efficient Entrepreneurship   8
   The Hardware Companies of Today   9
      Connected Devices   9
      Personal Sensor Devices (Wearables)   11
      Robotics   13
      Designed Products   14

2. **Idea Validation and Community Engagement**. . . 17
   Your Fellow Hardwarians   20
   Your Cofounder and Team   21
   Your Mentor(s)   23
   Your True Believers and Early Community   26

3. Knowing Your Market. 35
The Who, What, and Why of Your Product 36
Researching Your Market: Trends and Competition 36
    Market Size 37
    Market Trajectory 38
    Market Analysis 39
    Differentiators 41
Segmenting Your Market 43
    Customer Aquisition Cost (CAC) and Lifetime Value (LTV) 43
    Demographics and Psychographics 44
    Behavioral Segmentation 45
Customer Development 46

4. Branding. 53
Your Mission 58
Brand Identity and Personality 61
Brand Assets and Touchpoints 67
Positioning and Differentiation 74

5. Prototyping. 79
Reasons for Prototyping 79
Types of Prototyping 83
    Prototyping Terms 84
    Works-Like and Looks-Like Prototypes 85
    Teardowns 86
Assembling Your Team 87
    Industrial Design 87
    User Experience, Interface, and Interaction Design 88
    Mechanical and Electrical Engineering 89
    Software 90
Outsourcing Versus Insourcing 90
    Outsourcing 91
    Insourcing 91
Integrated Circuits 94
Connectivity 98
Software Platforms 102
Software Security and Privacy 107
Glossary of Terms 108
    Prototyping and Manufacturing Processes 108
    Electrical Components 110
    Sensors 111

6. Manufacturing........................... 113
Preparing to Manufacture 114
Where to Manufacture? 120
Supply Chain Management 128
Importing from Foreign Manufacturers 129
What to Look for During Manufacturing 131
Certification 134
Packaging 136
Sustaining Manufacturing 138

7. Acceleration........................... 139
Lemnos Labs 142
HAXLR8R 144
AlphaLab Gear 145
PCH 146
   Highway1 147
   PCH Access 148
Flextronics 149
Choosing an Incubator or Accelerator 150

8. Crowdfunding........................... 155
The Crowdfunding Ecosystem 155
   Kickstarter 156
   Indiegogo 157
   The DIY Approach 158
Planning Your Campaign 161
   Understanding Backers and Choosing Campaign Perks 161
   Pricing Your Perks 164
   Creating a Financial Model 169
   Timing with Manufacturing 171
Campaign Page Marketing Materials 172
Driving Traffic 175
   Leveraging Social Media and Email Lists 175
   Connecting with the Media 179
   Organizing PR Materials 180
While Your Campaign Is Live 183
   Data-Driven Crowdfunding and Real-Time Adaptation 183
   Publishing Updates for Your Community 186
Beyond Crowdfunding: Fundraising for a Company 187

9. Fundraising 189

First Things First	190
Bootstrapping, Debt, and Grants	191
Friends and Family	195
Angel Investors	196
The JOBS Act	197
AngelList	197
Venture Capital	201
Targeting Investors	201
Personalized Introductions	202
Telling a Story	204
Due Diligence	208
Strategics	209
Structuring Your Round	210

10. Going to Market 213

Business Models for Hardware Startups	214
Selling Additional (Physical) Products	215
Selling Services or Content	217
Selling Data	219
Open Source	219
Pricing	220
Cost-Plus Pricing: A Bottom-Up Approach	223
Market-Based Pricing: A Top-Down Approach	225
Value-Based Pricing: Segmentation meets Differentiation	226
Selling It: Marketing 101	228
Step 1: Define your Objective	231
Step 2: Choose your KPIs	232
Step 3: Identify Your Audience, the "Who"	233
Step 4: Select Your Marketing Channels	233
Step 5: Formulate Your Message	237
Step 6: Incorporate a Call to Action	237
Step 7: Specify a Timeline and Budget	238
Step 8: Refine Your Campaign	238
Distribution Channels and Related Marketing Strategies	241
Online Direct Sales	241
Online Specialty Retailers and Retail Aggregator Platforms	245
Small Retailers and Specialty Shops	248
Big-Box Retail	249
Warehousing and Fulfillment	261

11. Legal.	269
Company Formation	270
Trademarks	274
Trade Secrets	275
Patents	275
Manufacturing Concerns	281
Liability	281
Manufacturing Agreements	282
Import/Export Considerations	283
Regulatory Concerns and Certification	285
Medical Devices and the FDA	286
Hardware and the FCC	290
Epilogue: The Third Industrial Revolution.	291
Index.	293

Preface

This is a book about how to build a hardware business.

The goal of this book is to provide a roadmap for makers and hardware entrepreneurs who are looking to turn their product ideas into full-fledged businesses. Building a software startup has been largely templatized over the last five years. Free tools enable entrepreneurs to rapidly build, collaborate, deploy, and pivot. There is an extensive network of mentors and thought-leader bloggers sharing advice rooted in their own experience. There are bestselling books devoted to best practices, such as *Do More Faster, The Lean Startup,* and *From Concept to Consumer.* There are innumerable incubators and investors rounding out the ecosystem. It is now comparatively easy and inexpensive to launch a software startup.

Hardware is just starting to develop this type of ecosystem. Makers have been around for decades, but hardware startups have only recently begun to enjoy the type of attention and funding common among their software counterparts. Significant changes have taken place in the last two years. The decreasing cost of prototyping is lowering the barriers to entry and making it more feasible to develop a physical product within economic constraints. Kickstarter and Indiegogo have created platforms that enable fundraising for small-batch manufacturing and facilitate a community of early adopters. The number of hardware startups is on the rise. Venture capitalists who were previously uninterested in non-software investment are sitting up and taking notice. But the iconic how-to guide—the *Do More Faster* or *Lean Startup* of hardware—hasn't been written yet. Our goal is to be that guide.

Who This Book Is For

This book is meant for hobbyists and makers who are considering turning their idea or side project into something more.

It's for maker pros who have built a successful product and are interested in turning it into a company.

It's for the curious: software engineers considering switching from bits to atoms, or investors who want to learn more about this new ecosystem.

Basically, it's for anyone interested in understanding the challenges unique to founding and launching a hardware startup...and overcoming them.

How to Use This Book

The book is organized into the following chapters:

Chapter 1: The Hardware Startup Landscape
   The first chapter introduces the state of the market for hardware startups, dividing it into four main product categories: connected devices, wearables and personal sensors, robotics, and designed products. It briefly examines the forces that have led to the recent growth of the ecosystem, including the history of the maker movement.

Chapter 2: Idea Validation and Community Engagement
   This chapter begins by emphasizing the importance of validating the idea for your company through conversations with distinct groups of people who will be critical to your success as a company. It then introduces the topics of community building and customer development, discussing the different relationships that founders will form to help them along the path to building a company. These include the relationship between cofounders, how to choose advisors, and how to reach potential early adopters.

Chapter 3: Knowing Your Market
   This chapter covers techniques for market, consumer, and competitive-landscape research. It aims to help founders better understand where their products fit into a market ecosystem, which is important for idea validation, early brand positioning, and future fundraising. It also works through the basics of customer-development interviews with an eye toward lean product development.

Chapter 4: Branding
   This introduction to brand development for hardware startups covers the basics of brand marketing—including brand identity, mission,

and personality—and the development of brand assets. It will help founders craft their company's cohesive brand identity, which is a critical component of success for any physical product.

Chapter 5: Prototyping

This chapter is a guide for getting from design to physical *thing*. Topics include types of prototypes (including *works-like* and *looks-like* prototyping), building your engineering and design team, outsourcing versus insourcing, chip selection, software, and some common terminology specific to the hardware space.

Chapter 6: Manufacturing

This chapter discusses the common processes and pitfalls startups face when moving to manufacturing. It covers when and how to choose a factory and supply chain, where to manufacture, testing and certification, and packaging.

Chapter 7: Acceleration

This chapter presents a survey of the hardware-startup incubator and accelerator ecosystem. It covers the top programs supporting entrepreneurs today, with an eye toward differentiating their offerings and helping interested founders select the program that's right for them.

Chapter 8: Crowdfunding

Crowdfunding platforms have made a dramatic impact on the ability of hardware-startup founders to take an idea to market. This chapter covers best practices for running a crowdfunding campaign from start to finish: choosing perks, developing a pricing strategy, driving traffic, building community, and more.

Chapter 9: Fundraising

This chapter helps founders navigate the fundraising ecosystem. It examines the players who control capital—including *angels*, *venture capitalists*, and *strategic investors*—and the pros and cons of taking funding from each. It provides guidance on the strategies most likely to result in a successful fundraise, including when and how to reach out, how to create an ideal pitch deck, and how to structure a round.

Chapter 10: Going to Market

This chapter begins with a survey of business models and pricing strategies. It introduces logistics and fulfillment best practices and evaluates distribution channels, with special attention paid to margin and marketing considerations. It also covers the metrics that matter when evaluating the growth of a business. The emphasis throughout is on helping founders make the transition from *product* to *company*.

Chapter 11: Legal
Hardware-startup founders face unique legal considerations when building their products. They must navigate potential intellectual property issues, liability concerns, certifications, regulations, tariffs, supplier agreements, and more. This chapter provides an overview of the pitfalls to watch out for and the type of legal support a founder will need at various stages of product development.

Safari® Books Online

Safari Books Online is an on-demand digital library that delivers expert content in both book and video form from the world's leading authors in technology and business.

Technology professionals, software developers, web designers, and business and creative professionals use Safari Books Online as their primary resource for research, problem solving, learning, and certification training.

Safari Books Online offers a range of plans and pricing for enterprise, government, education, and individuals.

Members have access to thousands of books, training videos, and prepublication manuscripts in one fully searchable database from publishers like O'Reilly Media, Prentice Hall Professional, Addison-Wesley Professional, Microsoft Press, Sams, Que, Peachpit Press, Focal Press, Cisco Press, John Wiley & Sons, Syngress, Morgan Kaufmann, IBM Redbooks, Packt, Adobe Press, FT Press, Apress, Manning, New Riders, McGraw-Hill, Jones & Bartlett, Course Technology, and hundreds more. For more information about Safari Books Online, please visit us online.

How to Contact Us

Please address comments and questions concerning this book to the publisher:

O'Reilly Media, Inc.
1005 Gravenstein Highway North
Sebastopol, CA 95472
800-998-9938 (in the United States or Canada)
707-829-0515 (international or local)
707-829-0104 (fax)

We have a web page for this book, where we list errata, examples, and any additional information. You can access this page at *http://bit.ly/ the_hardware_startup*.

To comment or ask technical questions about this book, send email to *bookquestions@oreilly.com*.

For more information about our books, courses, conferences, and news, see our website at *http://www.oreilly.com*.

Find us on Facebook: *http://facebook.com/oreilly*

Follow us on Twitter: *http://twitter.com/oreillymedia*

Watch us on YouTube: *http://www.youtube.com/oreillymedia*

Acknowledgments

We would like to thank Chris Anderson, David Austin, Catherine Baecker, Marc Barros, Ayah Bdeir, David Bill, Chris Bruce, Jon Bruner, Jon Carver, Liam Casey, Claudia Ceniceros, Ben Corrado, Liza Daly, John Dimatos, Gene and Ursula DiResta, Dale Dougherty, Kate Drane, Ron Evans, Josh Fisher, Ash Fontana, Chris and Camille Forrest, Dan Goldwater, Marcus Gosling, Kaethe Henning, Justin Hileman, Xander Hileman, bunnie Huang, Firen Jones, Tom Kitt, Erik Klein, David Lang, Brian Lee, Jason Lemelson, David Lyons, Tim Mason, Christina Mercando, Eric Migicovsky, Sean Murphy, Tim O'Reilly, David Pendergast, Monisha Perkash, Ryan Petersen, Daniel Roberts, Jesse Robbins, Brian Sawyer, Jeff Schox, Jinal Shah, Dan Shapiro, Dave Shapiro, Andy Sherman, Eric Stutzenberger, Zach Supalla, Gnat Torkington, Jesse Vincent, John Vinyard, John W. Vinyard, Kathy Vinyard, Lizzie Vinyard Boyle, Sonny Vu, Saroya Whatley, and Kahlil Gabriel Williams.

CHAPTER 1

The Hardware Startup Landscape

IF YOU'RE READING THIS BOOK, IT'S LIKELY BECAUSE YOU'VE DECIDED TO start, or are thinking about starting, a hardware company. Congratulations! Launching a hardware startup is an exciting and challenging undertaking. There's a saying: "Hardware is hard." You have to navigate the complexities of prototyping and manufacturing, the daunting optimization problems of pricing and logistics, and the challenges of branding and marketing. And you'll be doing it all on a pretty tight budget.

But today—right now!—is probably the best time in history to be starting your company. Technological advances, economic experiments, and societal connections have facilitated the growth of an ecosystem that enables founders to launch hardware companies with fewer obstacles than ever before.

Before we get into the specifics of getting your business off the ground, let's set the stage by discussing some important influences that have brought the ecosystem to where it is today.

Early Makers

Today's hardware entrepreneurs stand on the shoulders of early makers. The maker movement has had a profound influence on the hardware-startup ecosystem. Defined by three characteristics—curiosity, creativity, and community—it emphasizes project-based learning, learning by doing, and sharing knowledge with others. Experimentation is important. Having fun is a priority.

While people have always had a desire to make things and work with their hands, the rise of a distinct hobbyist do-it-yourself (DIY) culture focused on technology began in the 1960s.

THE WHOLE EARTH CATALOG

Stewart Brand's *Whole Earth Catalog* (*http://www.wholeearth.com/index.php*), which first appeared in 1968, was one of the foundational resources of what became the maker movement. More than just a catalog, it was a manual for people who wanted to live a creative, DIY lifestyle, and a cornerstone of 1960s counterculture. Tools, machines, books, farming products—all of these could be found in the catalog, along with vendor names and prices. Customers could buy directly from manufacturers.

The catalog featured how-to guides on everything from welding to breeding worms. The emphasis was on personal skill development, independent education, and what's now called *life hacking*. John Markoff, technology writer for the *New York Times*, referred to it (*http://bit.ly/whole_earth_cat*) as "the internet before the internet" and "a web in newsprint." It captured the imaginations of a generation of counterculturalists, many of whom went on to careers in technology.

The catalog ceased regular publication in 1974 and was published intermittently until 1998. The back cover of the last regular edition had a farewell message: "Stay hungry. Stay foolish." This famous phrase is often attributed to Apple founder Steve Jobs, who called the *Whole Earth Catalog* "sort of like Google in paperback form" in his famous 2005 Stanford commencement address.

COMMUNITIES AROUND NEW TECHNOLOGY

In the 1970s, computers captivated the imaginations of many early Silicon Valley technologists, including Steve Jobs. Communities sprang up around the new technology. One example was the Homebrew Computer Club (*http://bit.ly/homebrew_cc*), a group of Silicon Valley engineers who were passionate about computers (particularly early kit computers). Members included Steve Wozniak and Steve Jobs. The club, which met from 1975 to 1986, was instrumental in the development of the personal computer. Wozniak gave away schematics of the Apple to members and demoed changes to the Apple II every two weeks.

These early adopters took the DIY ethos of the *Whole Earth Catalog* and extended it to DIWO (do it with others). At first, software was the primary beneficiary of this collaborative spirit. The free and open source software movements, which advocated the release of software whose source code was public and modifiable by anyone, began in the mid-1980s and steadily gained in popularity.

By the mid-1990s, the trend moved from bits to atoms, and an open source hardware movement began to grow (see "Open Source Hardware" on page 6 for more information and examples). Open source hardware (*http://www.oshwa.org/definition/*) is "hardware whose design is made publicly available so that anyone can study, modify, distribute, make, and sell the design or hardware based on that design."

MIT CENTER FOR BITS AND ATOMS

By the late 1990s, maker culture and prototyping technologies were becoming more formalized in academic institutions. Often called the "intellectual godfather of the maker movement," Neil Gershenfeld (*http://ng.cba.mit.edu/neil/bio.html*) founded the MIT Center for Bits and Atoms (CBA) in 2001. The CBA focuses on creating cross-disciplinary fabrication facilities that offer shared tools, with the intent to "break down boundaries between the digital and physical worlds."

These *fab labs* are scattered around the world, but they share core capabilities that allow people and projects to move freely among them. Projects (*http://fab.cba.mit.edu/about/faq/*) range from technological empowerment to local problem-solving to grassroots research. Several prominent companies with a distinct maker ethos and strong ties to the community have emerged from work done at the CBA, including Formlabs, Otherlab, Instructables, and ThingMagic.

MAKE MAGAZINE

As community-driven innovation and small-scale fabrication experiments were taking root in academia, the maker movement was steadily gaining popularity among hobbyists. DIY pursuits were steadily becoming more mainstream. Dale Dougherty, cofounder of O'Reilly Media and developer/publisher of the Global Network Navigator, noticed the increasing interest in physical DIY projects among his peers in the tech community.

Dale had previously created the *Hacks* series of books for O'Reilly. The series helped users explore and experiment with the software they used, empowering them to create shortcuts and useful tools. In 2005, Dale created *Make* magazine, based on a related, simple premise: "If you can mod software, you can mod the real world." He and O'Reilly Media cofounder Tim O'Reilly had spent years enabling people to learn the skills necessary to write software. *Make* was conceived to help them learn the skills needed to make things in the physical world.

In addition to teaching practical skills, *Make* emphasizes creativity. In 2006, the *Make* team put on the first Maker Faire (*http://makerfaire.com/*), with the goal of bringing the maker community together in person to showcase and celebrate the DIY spirit. Dale remembers:

> *I noticed that a lot of really interesting work was happening in private. We see objects every day, but there's nobody talking about how they were built. I wanted to create a place for people to have conversations about that in public, in a way that was enjoyable and fun.*

The 2006 event had 200 exhibits and drew 20,000 attendees. By 2013, there were 900 exhibits and 120,000+ attendees. The original flagship Maker Faires were held annually in San Mateo, New York City, and Detroit, and their numbers have exploded, with new Faires launching every year in Europe, Asia, and around the US. Community-run Mini Maker Faires (*http://makerfaire.com/mini/*) have popped up around the world as well.

As maker culture has become increasingly popular, thousands of people have been inspired to create unique projects that solve personal pain points or provide entertainment. Community hackerspace founders have taken the fab lab model and used it as inspiration for shared neighborhood workspaces. In addition, the rise of the Internet enabled communities to form unencumbered by geographical distance (see "Online Community" on page 7 for more information and examples). Technology enthusiasts from all of the world can connect with one another and share.

Technology Enables Scale

Over the past five years, we've begun to witness the emergence of *maker pros*: entrepreneurs who started out as hobbyists and now want to turn their creations into full-fledged companies.

The difference between a *project* and a *product* is the difference between making *one* and making *many*. To turn a project into a company, the product has to be scalable. "Making many" has traditionally been a problem of cost and accessibility; it's historically been both expensive and difficult to manufacture. Growing a company further requires keeping costs low enough to profit, setting up distribution channels, and managing fulfillment.

Over the past few years, several trends have combined to create an environment that's mitigated those problems. This has resulted in the growth of a hardware startup ecosystem.

RAPID PROTOTYPING

Advances in rapid prototyping technologies have fundamentally changed the process of taking an idea from paper to the physical world. Hobbyist and prosumer-level 3D printers, computer numerical control (CNC) routers, and laser cutters have altered the landscape of personal fabrication, enabling quick and affordable iteration.

While 3D printing has been around since the 1980s, the cost of a machine has dropped dramatically (*http://bit.ly/rapid_prototyping*). Materials such as metals and ceramics enable higher-fidelity models. Cloud-based fabbing services, such as Ponoko and Shapeways, can produce a single prototype and ship it to you within a week—no need to own the printer yourself!

Inexpensive boards (such as Arduino, Raspberry Pi, and BeagleBone) make electronics prototyping accessible to everyone. As interest in the Internet of Things has grown, products such as Spark Core and Electric Imp (startups themselves) have hit the market to make connected-device prototyping fast and easy.

Simultaneously, computer-aided design (CAD) software has become more sophisticated, more affordable, and easier to use.

INEXPENSIVE COMPONENTS

Just as the cost of major prototyping technologies has come down, component prices for sensors, batteries, and LEDs are also much lower. Several early maker businesses (MakerBot, Adafruit, SparkFun) are excellent resources for prototyping supplies and technologies.

Ubiquitous smart devices have also had a dramatic impact on the hardware ecosystem. Global smartphone penetration is at 22 percent; within the United States, it's at 56.5 percent and growing (and reaches as high as 70 percent in some countries). While this phenomenon is partially responsible for driving down the cost of component parts, the smartphone itself has also had a dramatic impact on hardware devices. It's an increasingly common interface through which humans can interact with connected devices and wearables.

SMALL-BATCH MANUFACTURING

As machine costs have dropped, small-batch manufacturing has become increasingly feasible. The minimum number of units needed to secure a contract manufacturer used to be in the tens of thousands, but today's factories are increasingly willing to do small-batch runs (sometimes in the hundreds of units).

Small batches are one way that a fledgling hardware company can run lean. Even if software-style constant incremental iteration is still impossible, the amount of money lost to a bad run is considerably reduced with a small batch. Increasing awareness of Shenzhen's growing manufacturing ecosystem and increased ease of sourcing through sites such as Alibaba and Taobao have also opened up China as a viable option for smaller startups.

OPEN SOURCE HARDWARE

Open source hardware platforms are continuing to gain popularity, allowing entrepreneurs to build on top of them. Arduino, for example, eliminates the need to build a proprietary board during the early development phase.

As of 2011, there were more than 300 open source hardware projects, and the number continues to increase. Engaged communities of contributors help accelerate innovation, and their accessibility and willingness to share knowledge draw in new makers. This democratizes innovation.

Open source can also be a business in itself; MakerBot and Arduino are thriving companies in their own right. By 2010, each already had over $1 million in revenue, and MakerBot was acquired by Stratasys in 2013 for $604 million.

The Open Source Hardware Association (OSHWA (*http://www.oshwa.org/*)) is the present-day voice of the open source hardware community. It works to advance the goals of collaborative learning and promoting the use of open source hardware.

ONLINE COMMUNITY

In addition to generating awareness of the hardware space and helping people learn more effectively, community knowledge-sharing helps spread best practices and innovative ideas. Web-based communities such as Instructables and Thingiverse are geography-agnostic; they enable people around the world to share projects online and learn from others. Sometimes communities come together to contribute funding to a particular project. Crowdfunding platforms help founders leverage community support and bring products to market.

While online communities can provide support and access to information, geographically concentrated local communities can help members overcome design and prototyping challenges by making access to expensive machinery much more feasible.

Hackerspaces (often called *makerspaces* if the primary focus is physical device hacking) are local hubs for hobbyists, crafters, and maker pros alike to come together and build things. Beyond creating a welcoming physical space and facilitating collaborative serendipity, they often include shared tools and machines (similar to the MIT Fab Lab model discussed in "MIT Center for Bits and Atoms" on page 3).

Some hackerspaces are simple community garages. Others, such as TechShop, involve paid memberships and offer courses for skill development. These spaces harness the power of sharing, creating an "access-not-ownership" model that makes even expensive professional-grade equipment relatively accessible. Over the past five years, makerspaces and hackerspaces have spread across the world. The Hackerspace wiki (*http://hackerspaces.org/wiki/*), which tracks spaces globally, lists over 1,600 spaces. Many are primarily devoted to software, but a steadily increasing number focus on hardware.

THE SUPPLEMENTAL ECOSYSTEM

Getting a device made is only the beginning of a successful hardware endeavor. To turn a project into a company still requires navigating fundraising, inventory management, distribution, customer service, and more.

New businesses are being launched specifically to help hardware startups navigate these challenges. Accelerators (Chapter 7) that traditionally offered funding, mentorship, and assistance to software companies are expanding into hardware. Hardware-specific programs are popping up, providing the specialized assistance necessary for startups to efficiently produce physical goods; some focus specifically on helping startups navigate manufacturing overseas. Fundraising platforms such as Kickstarter and Indiegogo can help entrepreneurs validate markets, raise money, and grow engaged communities (for more details, see "The Crowdfunding Ecosystem" on page 155).

Once the product has been made, fulfillment-as-a-service shops enable entrepreneurs to offload some of the logistical challenges of warehousing, packing, and shipping. Distribution channels, such as Grand St. (recently acquired by Etsy), Tindie, and ShopLocket provide a means to easily reach consumers without needing to go through big-box retailers.

THE "LEAN STARTUP" AND EFFICIENT ENTREPRENEURSHIP

The templatization of best practices for software startups has had a profound impact on entrepreneurship. The *Lean Startup* movement, introduced by Eric Ries in 2011, is a series of best practices designed to make starting a company a more feasible, less risky proposition for aspiring founders. Lean startups identify clear customer needs and incorporate customer feedback into the product design from day one, iterating rapidly to produce a truly useful product. They are strongly grounded in data-driven assessments of their offerings, using techniques such as A/B testing and closely monitoring actionable metrics (metrics that tie user data to outcomes). While running a lean *hardware* startup is fairly challenging, the popularity of the movement has inspired thousands of individuals to think seriously about turning their projects into companies.

One of the core principles of the Web 2.0 movement was that everyone is capable of "being a creator." Online, that spirit has been reflected in the rise of blogging, photo taking and sharing, pinning, tweeting, and creating web content. In the physical world, "being a creator" means making physical goods.

Dale Dougherty compares this progression from makers to entrepreneurs to a similar phenomenon that happened in the early days of the Web:

> Early on, most people were creating websites because they could. At some point, people said, hey, there's a way to make money from this—I'm not building websites; I'm building a way to make money.

The maker movement has increased the pervasiveness of the DIY spirit, facilitated easier access to information, and generated a supportive community that helps today's founders get companies off the ground. It has helped millions of people realize that they, too, can hack the physical world. People start off small: they learn how to make one of something. But once they've made one, making many—and starting a company—no longer seems like an impossible task.

The Hardware Companies of Today

Most hardware startups make products that fall into one of four subcategories: connected devices, personal sensor devices, robotics, and designed products. Admittedly, some hardware falls into multiple categories; your phone, for example, is a personal sensor device (it has an accelerometer and gyroscope that can be used to measure the activity of the person carrying it). When you open an app for your smart watch or fitness tracker, it becomes part of a connected device. And if paired with a product like Romo (*http://www.romotive.com/*), it can even become a robot.

Some products are difficult to classify, but for the purposes of this book, we've found that these categories make the most sense for discussing the challenges of bringing certain types of products to market.

CONNECTED DEVICES

The term *connected device* broadly refers to a device that has a cellular, WiFi, or other digital connection but is not a cell phone or personal computer. Some of these devices (ereaders, tablets) are designed to be used by people. However, the term is increasingly used to refer to devices that are connected to, and communicate with, other machines (machine-to-machine, or M2M). A growing number of connected-device hardware startups fall into this category. They are the startups that are building the Internet of Things.

The term *Internet of Things* was originally coined by Kevin Ashton (*http://bit.ly/that_iot_thing*), cofounder of the Auto-ID Center at MIT. Ashton recognized that most of the data on the Internet was gathered or created by humans:

> *Conventional diagrams of the Internet...leave out the most numerous and important routers of all—people. The problem is, people have limited time, attention, and accuracy—all of which means they are not very good at capturing data about things in the real world. And that's a big deal.*

The broad vision for the Internet of Things (*http://bit.ly/iot_mckinsey*) is a world in which objects connect to the Internet and transmit state information without human involvement. It's quickly becoming a reality. Cisco Systems, a manufacturer of networking equipment, states that as of 2010, there were 12.5 billion devices connected to the Internet (*http://bit.ly/iot_next_evolution*): "more things than people." By 2020, projections (*http://bit.ly/2020_connections*) from Cisco and Morgan Stanley estimate that 50–75 billion devices will be connected.

Internet of Things objects use networked sensors to generate data, which is then analyzed by other machines. The objects themselves can in turn be modified or controlled remotely. While many such devices have a consumer focus, the promise of the Internet of Things extends to Big Industry as well. Connected systems form the backbone of the Industrial Internet (*http://bit.ly/industrial_internet*), in which identifiers, sensors, and actuators work together to form complex autonomous systems in industries ranging from manufacturing to health care to power generation. The specific benefits vary—some industries are interested in cost reduction, while others care about improved safety—but the promise of the Internet of Things is one of better outcomes, improved by increased productivity and efficiency.

Making homes and cars "smarter" is a popular consumer vision. Startups such as Nest (recently acquired by Google), August, and Automatic are working on connected smoke alarms, thermostats, door locks, and vehicles. Others, such as SmartThings, are producing beautiful dashboards that act as a hub for monitoring connected home devices.

Given the vast market potential, many large companies have entered the IoT space. Belkin's WeMo plugs into electrical outlets and enables a smartphone to control the outlet (and the device plugged into it). In a new partnership with appliance maker Jarden, the WeMo Smart can turn Jarden's Crock-Pot and Mr. Coffee lines (among others) into connected and controllable devices. Lowe's produces the Iris Smart Home Management System, which offers sensors for security, temperature control, power management, and more, which all transmit information back to the owner's smartphone. In some cases, the human user need not be consulted; garden-soil monitoring sensors can detect dryness and automatically trigger the watering system.

Asset tracking is another popular application. Mount Sinai Hospital in New York has begun tracking assets (*http://bit.ly/tracking_assets*) (e.g., hospital beds, wheelchairs, and pain pumps) with radio-frequency identification (RFID) tags. Large farms are also getting in the game, marking cows with RFID tags to track when the animals feed and how much milk they produce.

This sector of the hardware ecosystem has benefited extensively from low-cost sensors and ubiquitous smartphones. Large-scale data-processing techniques have also had a profound impact; the ability to turn vast quantities of information into meaningful insights is increasingly important as more devices become connected. It's a great space to be building a business in. We'll be focusing on the pitfalls unique to hardware startups that make connected devices, such as security, standards, and power management. We'll also cover producing a seamlessly integrated software experience, ease of use, and competing with very large incumbents.

PERSONAL SENSOR DEVICES (WEARABLES)
The line between personal sensor devices and more general "connected devices" is somewhat blurry. The connected devices mentioned in the previous section are defined by their ability to autonomously communicate with other devices, and many wearables do exactly that.

For our purposes, "personal sensor devices" will refer to products that gather data related specifically to a human subject, then process and display it in a way that makes it easily understandable to human end users. This typically takes the form of a device (frequently worn by the subject) and a mobile app or dashboard that presents logs or visualizes trends in the data. A smart watch, for example, may track wearer step counts and sync to an app on the user's mobile phone.

The market for personal sensors and wearable devices grew organically out of the Quantified Self movement, which focuses on tracking personal data. As far back as the 1970s, people were experimenting with wearable sensors, but the movement began to gain mainstream attention around 2007. Gary Wolf and Kevin Kelly began featuring it in *Wired* magazine, and in 2010, Wolf spoke about it at TED. Since Quantified Self was a movement largely led by technologically savvy early adopters, it's not surprising that much of its early activity came from startups.

Health and wellness are the primary focus of most of today's wearable sensor devices. Activity monitors, which are designed to help people become more aware of their fitness practices, are the most common. Other wellness-device startups are attempting to tackle sleep tracking, weight monitoring, dental hygiene, and brainwave measurement.

On the diagnostics side, startups are pursuing blood glucose monitoring, smart thermometers, mats that alert diabetics to foot ulcers, and "smart pills" that monitor compliance. These applications portend a future in which the medical industry is increasingly reliant upon sensor technology. These devices often require some degree of US Food and Drug Administration (FDA) approval, which we will touch on briefly.

Outside of health and wellness, a broader wearables market has emerged with applications in fashion, gaming, augmented reality, lifelogging, and more.

The increased prevalence of smartphones, better battery life, low-energy Bluetooth connectivity, and the dropping cost of sensor production have made this an attractive space to build a company in. Widespread public adoption, particularly in wellness and fitness, has made the consumer market more attractive. Social networking and interconnectedness have driven user adoption, as friendly competition and data sharing help people set goals and stay motivated.

As in any growing space, big companies are paying more attention to the market in personal sensor devices. Nike has long offered run-tracking technology in the form of the Nike+ (a sensor that connects to a runner's shoe). In 2012 it expanded into the FuelBand, a bracelet that constantly monitors daily activity levels, though in early 2014 it announced that it was no longer producing the product. Reebok and MC10 have partnered to develop an impact-sensing skullcap for athletes in contact sports. UnderArmour's Armour39 is a chest strap and module (with an optional watch) that measure heart rate, calories burned, and general workout intensity.

These devices are a departure from the core business of these companies, but the combined temptation of a large addressable market (see the entry for "Addressable Market" under "Telling a Story" on page 204) and a desire to be seen as cutting-edge has led them to push the envelope. There's also the data; while the users benefit from it, the device manufacturer is learning as well...about the habits of its customers.

User experience and user interface design are particularly important considerations when building a personal sensor startup, and so is privacy. Data control is an issue that founders must keep in mind when building a business. We'll touch on these factors throughout the book.

ROBOTICS

The third subset of hardware startups is the robots. These automated machines are designed with an eye toward improving the lives of humans. Some, such as home cleaning robots or robotic "pets," just make everyday consumer life a bit more convenient or fun. Others are used for important tasks in industry, such as improving the efficiency of the assembly line or doing hazardous work such as defusing bombs. The wide-ranging applications and demand for robots spans the consumer, military, and commercial markets, making it an extremely lucrative space.

Autonomous robots are a relatively new technology, appearing in the second half of the twentieth century. The first, the Unimate (*http://bit.ly/robot_hof_unimate*), worked on a General Motors assembly line. Its job was to transport molten die castings into cooling liquid, and weld them to automobile frames. The Unimate was a large stationary box with a movable arm, but it did important work that was too dangerous for humans. Since the Unimate, robots have changed industry in three primary ways:

- Their accuracy, consistency, and precision have improved product quality.
- Their ability to do work that humans shouldn't, or can't, has made manufacturing safer.
- Their cost relative to value—particularly when factoring in increased productivity—has fundamentally altered the bottom line of many companies.

Robots now roll, balance, swim, sail, climb, and even fly (drones). Sensor technologies have enabled tactile, auditory, and visual input processing. Interaction experts are working to perfect the user experience of interfacing with robots. Improved actuator technology has reduced costs and size, and advances in computation have expanded the types of activities that robots can perform (and their degree of autonomy). It's a rapidly growing industry that continues to push the limits of what's possible in many other fields.

While large factory-floor robots are still primarily developed by big companies, there are a number of startups working in the industrial space. One example is the Baxter robot by Rethink Robotics (*http://www.rethinkrobotics.com/*), which is designed to work with humans on assembly lines without a safety cage.

Fetch Robotics (*http://fetchrobotics.com/*), founded in 2013 as Unbounded Robotics, is building a human-scale mobile manipulator robot at a $35,000–$50,000 price point, a fraction of the cost of similarly featured products.

Robots are gaining popularity off the factory floor too. They are increasingly being used in agriculture and farming; agricultural robots from Spanish startup Agrobot thin lettuce and pick strawberries. There are underwater robot (ROV, or remotely operated vehicle) startups such as OpenROV, and flying robot (unmanned aerial vehicle, or UAV) startups such as 3D Robotics and DroneDeploy.

In the consumer market, startups are tackling use cases ranging from telepresence to children's toys. Health care and home assistance are becoming popular sectors.

Like the hardware verticals mentioned previously, robotics has its own unique business and manufacturing challenges. Robotics products are often costly to manufacture, and finding the right partner can be particularly challenging. This can pose a challenge for a startup when it comes to identifying a go-to-market strategy, because it can be difficult to get the costs down to the point where the price is appealing to consumers.

Robotics startups have been particularly popular acquisition targets. In 2013, Google alone acquired eight robotics startups.

DESIGNED PRODUCTS

Our fourth category, designed products, is an admittedly broad catchall. These companies make purely physical devices; the startup is generally building hardware only, with no software. Some are simply *made things*; we're talking everything from 3D-printed custom dolls for kids to the kitchen gadgets sold in Bed Bath & Beyond.

A pioneering startup called Quirky changed the face of bringing a designed product to market by incorporating community. In Quirky's process, inventors submit ideas, and the community curates the ideas, provides feedback, and eventually votes on the market potential of ideas that gain a following. If the process goes favorably, the community continues to assist with research, design, and branding, and ultimately helps the inventor arrive at a final engineered product.

Following manufacturing (also handled by Quirky), the product moves into sales channels. This is a combination of direct and social sales, plus partnerships with retailers. The profit from the product is shared among the community as well, according to the impact each person made in getting it from thought to market. Quirky keeps 60 percent.

We chose to use a four-category framework in this book, because we'll occasionally point out business or engineering considerations that are specific to one particular subset. But generally speaking, the number of hardware startups across all categories has increased as the old barriers of large capital requirements and long lead times have largely fallen away.

So, to summarize: hardware is getting less hard. Increasingly available prototyping tools, such as 3D printers and laser cutters (which are themselves frequently the offerings of hardware startups), are enabling entrepreneurs to apply "lean" startup principles (see "The "Lean Startup" and Efficient Entrepreneurship" on page 8) to hardware companies. A growing ecosystem of support companies is helping to reduce the traditional complexities of marketing, inventory management, and supply chain logistics. We'll cover all of those factors in depth in this book, as we work through a roadmap for The Hardware Startup.

CHAPTER 2

Idea Validation and Community Engagement

One of the challenges of writing a book about how to start a hardware startup is that there isn't one best path to follow on the road to success. Therefore, each chapter in this book is designed to be a standalone introduction to a particular field of focus that is crucial for a hardware company's success. For instance, some founders will find that refining their prototypes early on makes the most sense. Other teams, particularly those in business-to-consumer (B2C) consumer electronics, will have to prioritize building a brand.

However, regardless of what sector you're building in, or what market you are targeting, *idea validation* is a necessity.

The process of founding a software startup was revolutionized by the introduction of the Lean Startup framework in the mid-2000s. Popularized by entrepreneurs Eric Ries and Steve Blank, the Lean methodology teaches founders to connect with potential customers and fully understand their needs before writing a single line of code. Blank used the expertise he gained founding eight startups to create and refine the *Customer Development* methodology, an approach to product development that focuses heavily on understanding the needs of the consumer.

Ries incorporated the Customer Development methodology into the Lean Startup framework. The *build-measure-learn* feedback loop (shown in Figure 2-1) illustrates the development cycle of a lean startup. The process relies on experimentation, iterative development, and understanding customer feedback with an eye toward building a product that consumers genuinely want. This is what Ries calls "validated learning."

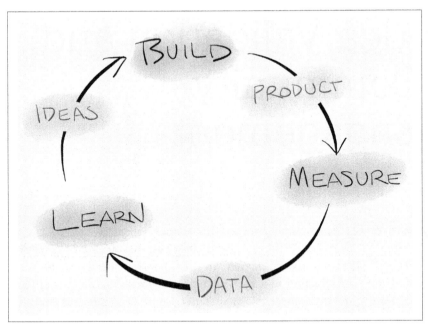

FIGURE 2-1. The build-measure-learn feedback loop

Founders typically decide to start a company because they have a desire to solve a specific problem, and a hypothesis about the best approach to do so. In the early days of a lean startup, the team refines their hypothesis through extensive customer interviews, with the goal of producing a *minimum viable product* (MVP). The MVP is a product that has the minimum feature set necessary to begin addressing the problem; further features will be added gradually, over time. This initial offering is typically tested on a supportive, carefully selected audience of early adopters who are capable of seeing the bigger vision even while the product is in the alpha stage.

Lean Startup development is *iterative*: the team continually tests the product within the target market, getting feedback through surveys and/or usage data, and making incremental refinements with each subsequent release. Gradually, more users are allowed into the alpha or beta test phase and they, in turn, facilitate a deeper understanding of customers' needs. The founders are simultaneously refining both the business model (how they make money) and the product experience.

The reality of building a hardware startup is that you can't do "lean" product development in the same way that a software-only team can. In an ideal world, you'd be able to follow the same iterative process: develop, test with an audience, and refine. However, it's extremely difficult to iterate on a physical product.

As prototyping technologies have improved, it's become increasingly possible to produce a basic version of the theoretical form factor of a given device: a design prototype. These shells are little more than models or 3D mockups, but they can give users a sense of what the product might look like. Using off-the-shelf boards and components, it's sometimes possible to prototype some of the MVP functionality in the early stages of development. In rare situations, a design prototype and functional prototype might even be the same piece. But in most cases, you won't have a whole lot to show your potential early customers.

The challenges of hardware don't end there. Hardware founders also have to be aware of regulatory issues, be familiar with the intellectual property (IP) and patents landscape, and understand the technological limitations of a potential design before they even approach a factory.

Lean Hardware may be challenging, but validated learning is still possible. In fact, the Lean Startup philosophy is partly based on the "lean manufacturing" (http://bit.ly/lean_mfg) process developed by Sakichi Toyoda of automaker Toyota. Toyoda's process emphasizes efficiency and prioritizes removing anything that does not add value to the final product. The Toyota Production System (http://bit.ly/muda_muri_mura) has three goals:

- Reduce waste (*muda* in Japanese) of all types: time, excessive inventory, overproduction, defects.
- Prevent overburden (*muri*) of people and equipment involved in production.
- Eliminate inconsistency (*mura*) in the production process.

Hardware founders should strive for hypothesis-driven development and validate their ideas with customers as early as possible. This can be done before or alongside the prototyping process, but it should absolutely be done before moving toward tooling. Changing your design or switching technologies becomes much more expensive once you've moved beyond a prototype, so it's important to get it right the first time.

The foundation of effective validated learning is *conversation*. You want to talk to—and, more important, listen to—several distinct groups of people who will be critical to your success as a company. Over the next two chapters, we're going to explore the validated learning process by exploring the different groups that will have an impact on your success, what they can help with, and how best to leverage their expertise. As a corollary to validated learning, we'll also cover market research and the types of questions you'll want to resolve before beginning to build your first prototype.

Your Fellow Hardwarians

The first group of helpers on the road to building your product are fellow hardware founders. Chapter 1 subdivided hardware startups into four categories: connected devices, wearables, robotics, and designed products. Presumably, you're already well versed in what's happening in your particular sector. If not, become familiar with the thought leaders and other founders in your space.

Not everyone within the hardware community is a potential customer who will help you unearth the roots of a particular pain point. But within this community are the people who have done it before. They have extensive experience and can provide invaluable guidance on specific problems. They'll help you reduce waste in the production process. Certain steps in the development process, such as finding a contract manufacturer, are often driven by word-of-mouth referrals. Networking with other founders building products in your space will give you a better chance of getting these things right the first time.

Many first-time hardware founders are coming from software, or are hobbyist makers who haven't developed extensive personal networks within the hardware-startup community. Fortunately, it's easier than ever to find fellow hardwarians. Online is the easiest place to start: the Hardware subreddit, LinkedIn groups, Facebook groups, and Twitter lists can help you stay on top of what's out there. Don't just search for "hardware"; get specific and look for "wearables," "Internet of Things," "industrial Internet," "sensors," "robotics," or whatever applies.

Although online communities can have strong ties, they aren't a substitute for a local network. Even if you live in a smaller town, meeting fellow hardware hackers nearby can help you discover facilities, suppliers, and other resources. These connections are particularly valuable early on, because you'll likely be prototyping locally. A great place to find folks in the "real world" is Meetup (*http://www.meetup.com*). Now there are hardware meetups happening in most major cities around the world!

The HackerspaceWiki (*http://hackerspaces.org/wiki/*) is another great resource for finding like-minded people, and it has the added benefit of potentially helping you discover a shared workspace or machine shop. See if there's a TechShop (*http://techshop.ws/*) or Fab Lab (*http://www.fabfoundation.org/fab-labs/*) in your area. In addition to teaching classes, these spots often host community events such as show-and-tells and happy hours. The Maker Map (*http://themakermap.com/*) can also help you discover what's happening locally; this site is maintained in part by the authors of this book, but it relies heavily on community contribution.

Your Cofounder and Team

If you already have a cofounder, or an early founding team: awesome! You're embarking on an exciting shared journey. If not, you're going to need to find one. It can take several months to convince someone to join your team, particularly if you haven't received funding, so start looking now. A cofounder relationship is like a marriage: you're going to be building a company with this person for years, so it's important to find an equally committed partner and establish good channels of communication right from the start.

If you're looking for a cofounder, consider signing up for a cofounder matching community, such as FounderDating (*http://founderdating.com/*), CoFoundersLab (*http://www.cofounderslab.com/*), or CollabFinder (*https://collabfinder.com/*). New communities like this are always launching, so do some Googling to find the best current options in your region or area of interest. If you want to do a hardware startup but don't have a specific idea that you are particularly passionate about, you can reach out to local startup accelerator programs about participating as an *entrepreneur in residence* (EIR) or *hacker for hire* (the latter is typically a viable option only if you have technical skills).

Building a well-rounded team right from the start is crucial. In the early days, this involves finding people you work well with, who have complementary skill sets. Know what you yourself bring to the table, as well as what you need.

Brady Forrest, coauthor of this book and head of PCH's hardware accelerator Highway1 (*http://highway1.io/*), describes the ideal early team for a hardware company as "a maker, a hacker, and a hustler." At a minimum, you need one founder with hardware experience and another with business experience (or a second technical founder who's willing to learn).

If you're technical, you probably have a good sense of what it takes to build a product. However, you might not understand sales, branding, or marketing very well, and at some point, someone needs to convince people to buy your device. Having a well-rounded business development person as an early team member can help ensure that you don't neglect this critical part of company-building.

If you're a nontechnical founder, you absolutely must recruit a technical partner. It will be virtually impossible to raise funding or efficiently get to market without one (in terms of both time and cost). A technical cofounder shouldn't be a hired gun you bring on to build your idea. You need to attract a teammate who shares your vision and wants to partner with you. Sales and finance experience isn't particularly useful in the early days of the company, so it's important to spec out specifically what you will be doing while your cofounder builds.

If you're planning to build a connected device, wearable, or any product that interfaces with your customer via software, you'll want to bring on a software engineer as early as possible, preferably one with some user interface (UI)/user experience (UX) design experience. Software is the arena where a hardware company can continue to innovate even after the physical product has been shipped, which is incredibly important. Some founders choose to outsource software development in the beginning, but you'll need to line up someone to own that critical product role fairly early on.

It's beneficial to build a pipeline of potential talent right from the start. Talent acquisition is yet another reason to get plugged into your local hardware community. Increasingly common in most cities, *hackathons* are a great low-pressure way to work on a project and meet people. If you went to a university with a strong alumni network, join your school's Facebook group, mailing list, or LinkedIn group.

Your LinkedIn connections are a powerful resource. The broader your network, the more likely you'll find potential teammates with the skill set you're looking for. Build relationships early. If you see someone you'd like to meet, ask for an introduction from a shared contact, even if you're not immediately hiring. Quora (http://www.quora.com) is also a great place to discover kindred spirits. Read questions pertaining to your space, and check out who's answering them.

As you begin to meet people, be open. Keeping your idea a secret won't get you very far. If you're passionate about your vision, talk about it with the potential teammates you're meeting. You need to know if their vision aligns with yours. Ideally, they'll bring their own ideas about the product to the table as well. When building a founding team, you're looking for collaborators, not people who will rely on you to make all of the decisions and drive 100 percent of the early development.

Once you've connected with potential teammates, work on some small projects to see if you mesh well together; founder incompatibility is one of the most common reasons young companies fail. You must discuss the questions of equity splits and division of labor as early as possible. Putting it off isn't going to make the conversation any easier.

You can find many great equity split calculators and blog posts (http://bit.ly/equity_split) to help you determine the appropriate initial equity allocation (http://bit.ly/founders_dilemma). Generally, considerations include the amount of cash fronted for early operations, prior IP contribution, and division of labor.

One common concern is *commitment*: are both/all cofounders quitting their jobs to do this full-time? Quitting a paying job before funding is secured is risky (and it's difficult to fundraise on just an idea), but a founder who does so is able to devote 100 percent of his time and attention to launching the new company. If one founder is working on the startup full-time while the other is still at another paying job, it is common for the founder(s) who took the risk to get more of the equity. In this regard, building a hardware company is no different than building a software company.

Your Mentor(s)

Another group of people to reach out to early in the idea stage is potential mentors. A good mentor will offer meaningful advice and help you work through challenges. She will typically have experience in some facet of your market or business. A mentor might have a particular area of expertise (like marketing), deep domain knowledge (for instance, a doctor who specializes in the ailment your device remedies), years of experience, or valuable industry connections. For a first-time hardware entrepreneur, building a close relationship with someone who has been through the process of manufacturing a product (and knows the pitfalls to avoid) is invaluable. Like fellow hardwarians, a mentor can also help you reduce waste and prevent overburden in your production process.

Reaching out to potential mentors can be daunting, particularly if you're doing so by cold email. People with a lot of experience are often in senior leadership roles and might seem inaccessible. For most of them, time is a valuable commodity in short supply. Therefore, your approach really matters. Telling someone you want him to be your mentor right off the bat is the wrong approach; it's like proposing marriage on the first date. Strong mentor relationships develop over time. When you reach out for the first time, follow these pointers:

- Keep the email or message concise and to the point. Long emails are often ignored, or the recipient postpones reading them until it's convenient.

- Tell the person why you're reaching out to her specifically, to avoid the impression that your email is a mass mailing.

- Be clear with your ask. Ask for help with a discrete question or clearly defined problem. Ideally, this should be something that justifies a personal response, rather than an issue he has already addressed in a blog post or other public source of information.

- Minimize friction by offering to accommodate her schedule and preferred mode of communication.

People who have made the effort to help someone generally enjoy hearing the outcome of the situation on which they gave advice. Be sure to follow up, both with an appreciative thank-you note shortly after the fact and with occasional updates as you continue to make progress. Some of these contacts will result only in single or sporadic conversations. Don't be discouraged. If you're working on something interesting, and reaching out to people who truly share your passion, you will find people who are excited about offering help and advice.

Some startup founders choose to formalize an advisory relationship and recognize a mentor's contribution by awarding a small equity stake in the company. This should involve a formal contract, in which the founder and mentor agree on the specific level of engagement that the mentor is expected to maintain. Monthly meetings or calls are considered standard. Strategic help, such as participation in recruiting or introductions to customers, or expert value-adds, such as work on a particular project, typically increase the amount of equity offered.

Don't make the mistake of choosing an advisor because he is famous and you want his picture on a slide to impress investors. If you're going to give away part of your company to this person, make sure he's doing enough work to merit the award. Be sure that your advisory contract includes criteria for termination, in case the relationship goes sour and the advisor doesn't deliver, or if you pivot into a space in which the advisor's expertise is no longer relevant.

While the equity structure of an advisory relationship varies across companies, Silicon Valley law firm Orrick put together the matrix shown in Table 2-1 (see its Founder Advisor Standard Template (*http://bit.ly/fastemplate*) for more details) to help founders make an offer in line with the market.

TABLE 2-1. Levels of company maturity

Advisor performance level	Idea stage	Startup stage	Growth stage
Standard: Monthly Meetings	0.25%	0.15%	0.10%
Strategic: Add Recruiting	0.50%	0.40%	0.30%
Expert: Add Contact & Projects	1.00%	0.80%	0.60%

Equity compensation is determined according to both the stage of the company (reflecting the risk of failure), and the degree of advisor involvement. The award is typically given in restricted stock or common stock options, and vests over one to two years.

In Chapter 7, we examine another form of structured mentorship: participation in an accelerator program. Most accelerator programs have an extensive roster of mentors with diverse backgrounds and skill sets. These groups of people are typically made up of founders, investors, and industry experts who have agreed to help accelerator-backed teams.

Participating in an accelerator program is often a great way for a first-time entrepreneur to gain access to a valuable network.

Your True Believers and Early Community

Your True Believers are your earliest evangelists. They're the people who care enough to actively help you on the road to success, right from the start. Our favorite definition of True Believers comes from David Lang, cofounder of OpenROV, in his book *Zero to Maker* (Maker Media, Inc.):

> A True Believer is someone who knows you, the person behind the art or product. Someone you've confided in by showing them your art or business plan. They care about your product, but they also care about you. Not only will they buy your product, but they'll tell everyone they know about what you're doing; they'll get the word out.

The core of this group is usually your personal friends and family. Mentors and close connections you make in the hardware community can become True Believers as well.

Gathering your True Believers requires taking a close look at your personal network. Start small: tell your closest friends and family what you're doing and ask if they'd be interested in regular updates. Then, widen the circle. If you already have a following on social media channels, leverage it: tweet, post, drive friends and followers to a landing page with a forum or email list sign-up. Identify the influencers in your network who will likely be interested in your project, and reach out to them individually with a dedicated note. Tell them what you're doing and why you're excited about it.

There's no magic ideal number of True Believers. However, since one of the functions of this group is to help you get the word out, increasing the number of True Believers can increase your reach, so it's worth spending time cultivating these relationships. The more engaged this group is, the easier it will be to successfully run a crowdfunding or preorder campaign down the road. You should start to involve the people with whom you have a strong relationship as early as possible, even if you don't have much to share at first.

True Believers are a subset of a bigger group: your early-adopter community. Communities typically form around interests, practices (for instance, hobbies or professions), or location (geographic proximity). Sometimes, they are working toward a shared goal and can be considered communities of action. Or, alternatively, they might be bound by shared circumstances, such as life stage. It's important to [identify _why_ your earliest adopters should want to spend their time participating in your community.] What will they get out of the experience? They are probably interested in your product, but they won't want to spend a big chunk of their free time endlessly discussing something that doesn't exist yet. While True Believers care about _you_, the slightly broader early-adopter community generally cares deeply about _your space_. As a result, they're an excellent source of unvarnished feedback for early idea validation.

Getting a new community off the ground can feel like an insurmountable obstacle; it's really hard to build a group from nothing. To grow your community beyond the True Believers who love you, you'll have to put some effort into extending and strengthening your network. If you don't have a strong circle of contacts, or have avoided developing a presence on popular social channels, consider putting in a small amount of time each day to foster one. You don't have to become a social media guru, but having a rich network can only help you.

Establish relationships by participating in online groups. Post interesting content on Twitter, and engage with people who are commenting or sharing articles about your space. Google a wide variety of keywords and phrases, see what blogs or sites exist, and reach out to the authors. Look for in-person meetups. If you don't find any, set one up. You might find like minds in academia as well; for example, many universities have robotics and other hardware-focused clubs.

Your goal is to find passionate people who are as excited as you are about the problem you're solving. As David Lang of OpenROV puts it, "You're building a product that you really want. Sort through the other 7 billion people on the planet, and find the others who are like you." (Find out more about David's advice for starting a new community in "OpenROV: A Case Study" on page 27.)

OpenROV: A Case Study

David Lang, TED Fellow and author of _Zero to Maker_, is one of the founders of OpenROV, a Bay Area startup building underwater remotely operated vehicles (ROVs). OpenROV's open source robots are designed to make underwater exploration accessible to everyone. Right from the start, David and his cofounder Eric Stackpole envisioned a robust community of underwater-exploration aficionados helping to move the project forward. Here, David shares his experience with building a community from the ground up.

OpenROV was founded in 2010. David and Eric wanted to explore an underwater cave, but the equipment necessary was too expensive and they didn't have a research grant. So, they decided to make their own ROV. When David and Eric began to develop the robot, "it didn't work," David says. The technology simply wasn't there yet. Raspberry Pi and other single-board Linux computers hadn't hit the market.

"We tried to skate to where the puck was going," David says. They experimented with different solutions and reached out to others who were working on projects with similar technical challenges. They decided to make OpenROV an open source endeavor. By creating a project with strong ties to the maker community, they would be able to learn from others who were passionate about both technology and underwater exploration.

At first, there were no others. They set up a Ning (*http://www.ning.com/*) site, but for the first year, only David and Eric posted in the forums. "Until you have hardware out there that people are actually playing with, it's very, very hard to foster community," David says. People need to touch the product before they get excited about it. The team found that one way to bridge that preproduct gap is to work on creating a really interesting narrative around the problem and the product. They worked on telling their story in a way that piqued interest and got people excited about the problem and the team until the product was ready to touch.

"I think a lot of businesses don't have a good enough narrative around what they're doing," David says. He and Eric prioritized storytelling from the start. They used blogs, interviews, and social channels to share their vision of OpenROV as a tool for exploration and education for all. David wrote about his own progression from total newbie to maker pro and startup founder in the "Zero to Maker" column on Makezine.com. The team regularly posted on the OpenROV blog and forum. Over time, users began to find them.

David likens early community building to starting a fire. In the beginning, your setup is important. You only have kindling. "You need to create a spark in order to get things going, then build up to a healthy fire," he says. Foster the spark by providing it with material to keep it continually growing. The spark must be compelling. In OpenROV's case, they emphasized a vision: "people working together to create more accessible, affordable, and awesome tools for underwater exploration." This clearly articulated vision resonated with their early fans. Makers were drawn to the project, which resulted in a product shaped by the community. Open source contributors actively participated in the hardware development. "One of our users designed a new electronics board," David says, while "someone else worked on how the camera moves inside the tube. Another prototyped a new battery."

As the project attracted the attention of a broader audience, forum participation picked up. "I still read every single post in our forums. We work really hard to set the tone, make people feel comfortable asking questions," David says. He pays close attention to the most active contributors. If someone new joins and is very active, David will reach out and have a Google Hangout with them. Sometimes the team will send out free hardware in order to keep the most active users contributing feedback (or code). Occasionally, community members accompany OpenROV on an expedition. The team's first expedition was a trip down the California coast, around Baja, and into the Sea of Cortez, and a few passionate community members joined them on the boat. "The most important thing for us is that people who interact with OpenROV are excited to come back," David says. OpenROV regularly exhibits its robot at Maker Faires and other maker- and education-focused events. Being there in person increases their visibility and helps them reach a different audience.

As a community has taken shape, the OpenROV narrative has expanded. + and David recently created OpenExplorer (*https://openexplorer.com/*), a site that documents expeditions and excursions and encourages other adventurers to share their own exploration plans. This site isn't limited to underwater exploration; its goal is to foster community among explorers and citizen scientists of all kinds. It's an extension of the open, collaborative vision that has fueled the project from the start.

The OpenROV team didn't approach community building by creating a prototype and then trying to grow a community of prospective buyers. Instead, they articulated a well-defined vision and mission that attracted kindred spirits, involved them in the project, and kept them engaged.

Once you've identified a handful of potential early adopters, it's time to facilitate dialogue and begin turning the group into a true community. Introduce the members to one another and seed a conversation. You're going to have to do the work of keeping conversations going until there are enough engaged users that it happens organically. Tailor your early content accordingly. To the extent that you're comfortable, talk about the development process you're going through, and solicit feedback (again:

idea validation!). Passionate community members will enjoy feeling like they're part of the process. You should try to maintain a level of excitement that will keep them evangelizing for you through what is potentially a long path to market.

Work hard to avoid making your community discussions all about you or your product, because that will get boring quickly. Discuss the broader industry or space. Share news articles about the underlying issues you all care about. Call attention to the milestones or successes of the other members of the group. Host a local meetup and socialize in person over drinks or pizza. Above all, make your True Believers and early adopters feel valued and heard. Show appreciation throughout your path to launch. Consider offering special perks to show appreciation, particularly as you begin to take preorders or ramp up to a crowdfunding campaign.

To keep the group growing, encourage early members to refer trusted contacts from their own networks. Reach out to your top community members individually to ask if there's anyone you should invite to participate in the group. Having an invitation-only model in the beginning can help you control the size and tone, as well as establish community norms. As the group grows and you no longer have to personally drive engagement, you can make it public and open and begin to promote it.

It's easy to host your group using one of the many free tools available online. In the early days, a mailing list can get things started, but email lists don't scale very well; setting up a Google or Facebook group is quick and effortless. Creating a Ning or blog-based forum requires more work, but it's more customizable. It's also likely to rank high in search results, which can help people find you.

Chris Anderson, founder of 3D Robotics (*http://3drobotics.com/*), set up the initial DIY Drones (*http://diydrones.ning.com/*) community on Ning. He was able to find a cofounder and grow an early-adopter community on the site (see "3D Robotics: A Case Study" on page 31 for more details). If you have a project that will appeal to a very technical or maker-oriented audience, consider creating a forum on Instructables (*http://www.instructables.com/*). Whatever platform you choose, make sure it's a format that your community is receptive to. Once the community is public, be sure to facilitate easy sharing of content to other networks, so that others can discover you.

3D Robotics: A Case Study

Chris Anderson is the founder of 3D Robotics, an unmanned aerial vehicle (UAV) startup based in San Diego and Mexico. Here, he shares the ways that observing a community of engaged users shaped the company.

Chris discovered his passion for UAVs in 2007, while playing with LEGO Mindstorms with his children. The kids were bored with the out-of-the-box experience, so Chris Googled "flying robots." That search led them to combine some sensors and a model airplane with a Mindstorms set, and to write some software. "We did it just right enough that it flew," Chris says. "The kids lost interest, but I got chills. There on the dining room table with toys, we built something that the government regulates. I had a gut feeling that something was going to be different, that there was a tectonic shift in where the world was going."

Chris set up a Ning social network (*http://diydrones.ning.com/*) and began to post questions to learn more about UAVs. He met his cofounder, Jordi Muñoz, through the site. Jordi had felt similar chills when he discovered that he could control a toy helicopter with an Arduino. Since there was so little information on the Web at the time, the site quickly became the premier hub for personal UAV enthusiasts around the world.

While they credit much of the early success of DIY Drones to being the first mover in a new market (right place, right time), Chris and Jordi made a smart strategic decision to model their new community after a successful existing platform: Arduino, the open source microcontroller. Arduino's model— a strong focus on community and a simple, easy-to-use platform—was an inspiration for Chris and Jordi. "Arduino succeeded because it was simple and easy and everyone wanted it," Chris says. "It had Web in its DNA, community in its DNA." Chris and Jordi embraced the popularity and simplicity of Arduino, leveraging its brand in their own designs: ArduPilot, ArduCopter, and ArduPlane.

3D Robotics grew out of the DIY Drones community into a standalone company in large part due to user feedback. Chris noticed that while some participants on DIY Drones were interested in the full DIY experience from day one, a large number of them wanted to start exploring UAVs by buying a kit or ready-made drone. Observing user demand inspired the team to create a business around the hobbyist community.

Paying attention to community activity also shaped their business model. While the team acknowledges the influence that open hardware had on the early vision for DIY Drones, they noticed that their users weren't actually participating in open hardware development:

> No one out there is modifying the EAGLE files and giving them back, or exchanging CAD files. But activity on the software side is incredibly robust. As a company, you've got to know what you are. Over time, it became clear that what we are is [akin to] the Android operating system—not the phone.

As a result of these signals, the 3D Robotics team began to deemphasize open hardware.

Engaged users continue to shape what 3D Robotics offers. The team pays particularly close attention to new and nontechnical users. "They are the most active; they're asking for the most help," Chris says. "You could argue that they are the noise, but I think in fact that noise is a signal for us. Everyone asking for help is identifying a design flaw." Chris still personally reads the forums and answers technical support questions. He believes that understanding the problems of confused users ultimately provides the team with insight into how to make the user experience better, and how to make the product simpler. "It's unvarnished feedback," he says.

While DIY Drones and 3D Robotics undoubtedly benefited from first-mover advantage when it came to growing their initial community, the important takeaway is the power of continually engaging users. "Fundamentally, we are a user community," Chris says. "A not-insignificant portion of our users become contributors, even if all they do is help with documentation. We have a culture of participation."

Carefully observing a passionate user community can help a founder make smarter decisions about everything from the design of the product to the user experience to the structure of the business itself. Community activity may help an entrepreneur validate a hypothesis about her product without the need for a formal survey or focus group. It's worth a founder's time to pay attention to why the community is interested in the product, and to proactively foster engagement.

The approach described in this chapter of building communities with an eye toward gathering feedback and testing hypotheses is largely geared toward founders creating consumer-focused products. However, it's possible (and useful!) for business-to-business (B2B) oriented founders to approach the process in a similar fashion. Connecting with the local hardware community, finding mentors, and identifying a true partner to cofound your company are important for exactly the same reasons. While B2B startups are less likely to be able to leverage friends and family when building enterprise hardware, building a trusted network of potential early adopters is valuable.

Employees of companies that might become your customers will be a source of extremely valuable product feedback. While you might not want to reach out to the C-suite in the earliest stages of development, finding potential early adopters is still important. If you have clearly defined end users within a business, creating a community of potential evangelists is extremely worthwhile. If they love the product concept, they might try to sell you into their company. Software startups such as Twilio, Stripe, and GitHub have done an exceptional job of leveraging the developer community to make inroads into enterprise.

CHAPTER 3

Knowing Your Market

Your mentors, True Believers, early adopters, and the broader hardware-founder community will all provide invaluable feedback in the early stages of idea validation for your product. However, since your goal is to build a profitable company, one group of people is clearly the most important: the customers. They're the people who will give you dollars. You can't have an individual relationship with all of them, unlike the groups discussed in Chapter 2. However, you must know everything you can about them—who they are, what they need, what drives them, how much they're willing to spend—so it's extremely important to have conversations with them early in the product development process.

The time to identify and talk to potential customers is long before you're thinking about moving toward manufacturing. The lessons you learn from these discussions will shape the product. They'll also be extremely important for branding and marketing strategies, setting a price point, and identifying the distribution channels that are right for your company.

In the preprototype stage of customer development, the primary question you're trying to answer is, "Who is likely to buy my product?" You can't be all things to all people, so it's important to identify the type of customer who is most likely to spend money on the first version of your widget. You'll revise your offering for mass mainstream adoption somewhere down the line.

The Who, What, and Why of Your Product

Building any company—hardware or software—will require you to understand your market, know your customer, and build your community. Your goal is to find both problem/solution fit and product/market fit. In his book *Running Lean* (O'Reilly), Ash Maurya defines problem/solution fit as "Do I have a problem worth solving?" Product/market fit is "Have I built something people want?"

Before you build a single prototype, you should consider three important questions:

- What is the problem I am trying to solve?
- Who are the people who have that problem?
- Why should they want to buy my product rather than the solutions that already exist in the market?

The purpose of these questions is to help you identify and segment your market, as well as to understand the underlying drivers that lead customers to purchase products in your space. The research you do as part of the idea-refinement process is extremely important. It will shape your early thinking on technology and pricing, and it might have an impact on fundraising strategies in the future. It will also help you begin to define your brand.

Researching Your Market: Trends and Competition

When it comes to gathering numbers, market research for a product idea can be broken down into an examination of *market size* and a forecast of *market trajectory*. The size is a function of sales revenue and users over a given time period (possibly also within a geographically constrained area).

For example, to determine a rough estimate for the market size of a given medical device, you would multiply the number of potential buyers in the market (hospitals, clinics, etc.) by the quantity of devices each would purchase in a year, and then multiply that number by the average price paid for a single device:

> 5,723 hospitals in the US (*http://bit.ly/hospital_facts*) × 10 devices purchased per hospital × $3,000 cost per device = $171.69 million

This example counts hospitals, but it neglects to consider other buyers, such as clinics. However, it also assumes that every hospital in the US would buy 10 devices, which is unlikely.

MARKET SIZE

One common pitfall for new entrepreneurs is overestimating the number of potential buyers. If you are selling a device targeting hospitals, your potential addressable market is not the total number of hospitals in the world. It's the number of hospitals that fall into your potential sales channels (perhaps limited by geography or business structure) *and* are likely to actually buy your product. Some might be buying from your competition, or perhaps they don't want (or need) your device. In order to be more precise, marketers therefore use three distinct levels when referring to market size (nested as shown in Figure 3-1):

Total addressable market (TAM)
> Sometimes called the *total available market*, TAM is an estimate of the maximum potential revenue opportunity for a given product or service. Assuming no competition and no distribution challenges, this is everyone you could possibly sell to in an ideal world. It's important to clarify the geographic scope of TAM; some people assume it's global, while others identify a particular region.

Serviceable available market (SAM)
> A subset of the total addressable market, the term *serviceable* identifies customers whose needs are served by a specific product offering. SAM market sizing accounts for the fact that a given market has competition and that companies are limited by distribution channels.

Serviceable obtainable market (SOM)
> SOM is the realistically obtainable market, limited by factors such as competition, cost, outreach required, distribution channels, and so on. These are the customers you have a realistic chance at closing.

In our hospital example, the TAM might be all of the hospitals in the world. SAM would impose some limitations: the hospitals must be located within the US, must have more than 20 rooms on the cardiac floor, or must be running a particular operating system. SOM would further limit the market by imposing likelihood constraints: if 20 percent of hospitals are in a long-term contract with a competitor, they are likely not obtainable. When considering how big you can potentially grow a company, the most relevant estimate is almost always the SOM.

FIGURE 3-1. TAM/SAM/SOM market levels

MARKET TRAJECTORY

Market-size calculations give you a snapshot of overall market potential at a given moment in time. The market *trajectory* refers to whether a given market is expanding or contracting. It's about trends. A market can grow or shrink as a function of many different factors. In the previous example, perhaps those 5,723 hospitals are facing budget cuts and 5 percent of them will disappear in the next year (this assumption is totally fictional). If that year-over-year (YOY) trend continues, perhaps selling to hospitals isn't a great long-term play, and it's time to consider selling to clinics instead.

For consumer products, demographic changes are an important factor in predicting market trajectory. One commonly cited example is the number of senior citizens (age 65 or older) in the US, which is rapidly increasing (*http://bit.ly/senior_pop_trends*). There were 40 million seniors in the US in 2010 (13 percent of the population), but projections indicate that this number will grow to 72 million (20 percent of the US population) by 2030. This bit of information doesn't tell you anything about the purchasing habits of this demographic, but since the population is growing, it is reasonable to expect demand for products serving it (including health-related devices, home-assistance tools, entertainment platforms, etc.) to grow as well.

At other times, the number of potential users may remain constant, but shifts in industry business practices might lead to market expansion or contraction by way of changing purchasing habits. For example, one of the reasons that BlackBerry's market share declined substantially is that companies began to allow employees to use their own mobile devices at work.

Gathering data on numbers and trends can be somewhat challenging. Google searches surface news articles or blog posts that mention market size, but those posts are often pulling their data from market research firms' white papers. The white papers themselves are rich sources of information, but they can cost hundreds to thousands of dollars. Occasionally, you can gain access by signing up for a trial. Quora is a good place to go to see if others have posted relevant data, or to ask a question of an industry expert yourself. And Steve Blank, one of the creators of the Lean Startup, maintains a market research section on his blog (*http://bit.ly/blank_mrkt_research*) that links to open data sets and helpful market-sizing resources.

It's less important to have exact numbers than it is to have logical justifications for believing that a market you're looking to enter can support your potential company. Running analyses based on several different scenarios (called a *sensitivity analysis*) can help you evaluate the degree to which a specific assumption weighs on a given outcome. When doing back-of-the-envelope estimates of market size, run the numbers based on best case, worst case, and expected outcome.

MARKET ANALYSIS
The overall dollar value of the market, the number of potential customers, and the trajectory are all important variables, but market research is about more than just quantitative facts and statistics. We started this book with a look at the factors that have combined to drive growth in the hardware startup "industry" (decreasing cost of components, better and faster

prototyping technology, etc.). The goal of such research is to understand the interplay between forces shaping your market. Studying the trends underlying a market will help you understand your customer, develop effective brand positioning, and, eventually, communicate the potential of your idea to an investor.

Venture capitalist and Hunch cofounder Chris Dixon calls this type of market sizing analysis "using narratives, not numbers." (*http://bit.ly/ narratives_not_numbers*) To understand your market in narrative terms, consider economic, social, political, and technological factors driving its growth. For example, according to industry analysts (*http://bit.ly/ data_from_iot*), more data will be generated by machines than people next year. The factors driving this phenomenon include advances in analytics tools for processing big data, cloud computing (no need to buy machines to dedicate to analysis), and ubiquitous connectivity. Combined, these trends indicate that demand for enterprise Internet of Things hardware is likely to increase.

Once you have a handle on market size and trends shaping your sector, it's time to look at the specific companies and brands that operate within it. One of the most important things to determine is how crowded your market is. More-established markets typically have more existing players, including some big brands.

Consider smart watches. The first watches appeared in the market approximately 10 years ago and were primarily for receiving time-sensitive information (weather, stock quotes, etc). As the technology developed, fitness-tracking features were added, which increased the appeal to athletes. Over time, these devices have added features—GPS, ability to receive text messages and email, heart rate monitoring, etc.—and become increasingly useful to a mainstream audience. There are currently dozens of watches on Best Buy's Smart Watch page (and there are undoubtedly more that Best Buy doesn't carry). Most are from large companies, including Samsung and Sony. But there are also offerings from startups such as Pebble.

There isn't a specific number of companies that separates a "crowded market" from one that's still developing. What matters is the combination of how many companies are in the market, how strong their brands are, and how much money is being spent in the space. If you're going to compete in a market sector with many incumbents, knowing the players involved will help you identify and prioritize your differentiators.

DIFFERENTIATORS

Differentiators are what distinguish you from your competition. They are the answer to the question posed earlier: "why would a customer buy my product over my competitor's offering?" Particularly in a more mature market, some of the products you're competing with belong to companies with millions of dollars to spend on advertising and physical showrooms already in place. Startups can compete with incumbent giants, but it's critical that you determine in advance if what you'd like to build is different enough to claim meaningful market share. If you're producing an activity-tracking product to compete with Jawbone Up, for example, will your product compete on price (by being less expensive)? Will it offer more features? Perhaps it will be more accurate or have a more stylish design?

The most basic way to get a handle on your competition is to identify which products exist and make a *feature comparison matrix* (see Figure 3-2); this mimics the process that consumers go through when they buy devices (particularly if they're shopping online). A good way to start is to do extensive online research on as many product-related keywords or phrases as you can think of. In addition to unearthing competitive products, it will give you a preliminary sense of how the products are marketed, what sites (and physical stores) are selling them, and what publications or blogs are reviewing them. Once you have compiled a list of products, go looking for reviews (Amazon.com can be a particularly useful site to check). Read all of the reviews, and make a note of the features or usability factors that customers are praising or complaining about. This is an excellent source of unvarnished feedback.

Try not to let your research happen entirely online. You, and your competitors, are building physical products. A feature matrix can tell you only so much. It can't convey how pleasant the user experience is, or if the components have a high-quality feel. Get out there and touch the competition's offerings, to the greatest extent possible.

FIGURE 3-2. A feature comparison matrix

Differentiators aren't limited to cost, design, and features. You can also appeal to convenience, quality, or specifically targeted use cases. You can offer better customer service or a more intimate or positive customer experience. The important thing is to have a solid understanding of what already exists and why people are buying it. This will help you identify gaps in the market—unmet needs that you can set out to fulfill. As you build your brand, this early knowledge of your differentiators will help you to position yourself effectively.

If you're not competing in an established market, you have a different set of problems to consider. First and foremost: does anyone really want your widget? If you truly have no competition, is it because no one really cares about the "problem" you're trying to solve? Creating technology in search of a problem might mean that the timing isn't right for your offering. You will not only have to explain your product to an uneducated audience, but you'll also have to convince them that they should pay you money for a solution to a problem they didn't know they had.

On the flip side, however, you might truly be onto something big and new. Lots of successful companies were founded to meet the needs of what initially seemed like a niche market. Jawbone, for example, started as a company that built noise-canceling headsets for US soldiers. The thought that the technology might appeal more broadly to consumers came several years after initial product development. The "connected home" space is another relatively recent transition from niche hobbyist community into major consumer market. If your product is built upon a real technological breakthrough, there simply might not be a market out there yet.

Chapter 10 goes deeper into how to leverage your differentiators to build a distinct brand, as well as marketing within both crowded and non-established markets. For now, at the preprototype phase, you should be looking at the data you uncover with a critical eye and with the goal of deciding whether the market you're considering entering is big enough for the type of company you're envisioning building.

Segmenting Your Market

Market segmentation is the process that most marketers and brand developers use to identify groups that are most likely to receive a new product positively. Segmentation involves dividing a broad potential market into distinct subsets of people who share common characteristics or needs. The customers within those subsets will likely share a similar response to particular media channels or advertising approaches. Once the brand has identified relevant segment attributes, it can design a distinct marketing mix most likely to reach them.

The goal of segmentation in the preprototype phase is to help you discover and understand who you're building for, so that you can get both your product features and brand positioning off on the right foot from the very beginning. You can't build an MVP without knowing whom you're building for.

Segmentation helps you identify and focus your energy on the customers who are most likely to buy from you. It's a way to identify potential early adopters as well as customers to reach out to later in your company's lifetime. An ideal target segment is distinct enough to be measurable and also large. It should be stable (meaning it will be around and relatively unchanged for a while) or growing, and it should have enough purchasing power to enable you to turn a profit.

CUSTOMER AQUISITION COST (CAC) AND LIFETIME VALUE (LTV)

As discussed in "Market Size" on page 37, a meaningful target segment must be *accessible*, which means that you can reach out and market your product to that audience. Software startups, particularly those practicing the Lean Startup methodology, typically keep track of two metrics related to customer acquisition from the very start. The first is *customer acquisition cost* (CAC). This is the amount of money you spend to get a customer to buy your product or try your service. It includes money spent on market research, advertising, promotions ("$10 off your first order"), and sales (including the salaries of your sales force).

The second is *lifetime value* (LTV), which is the amount of money you will theoretically make from a given customer over the duration of his relationship with your company. Chapter 10 spends a lot more time discussing CAC and LTV in the context of distribution channels, but even in the early, preprototype days, it's important to develop a sense of what it might cost to reach your customers. If they don't pay attention to traditionally inexpensive marketing channels, or if you can't reach them without extensive effort or cost (in terms of time or money), they might not truly be a viable market segment for your product, particularly when you're just starting out.

DEMOGRAPHICS AND PSYCHOGRAPHICS

Markets can be segmented according to many different factors. Two of the most common are *user characteristics* and *user behaviors*. Segmenting a market on the basis of customer characteristics involves two types of customer profiling:

Demographics
   Demographics are quantifiable statistics that describe a particular population. A marketer's basic demographic profile might include age, race or ethnicity, and gender. Other demographic data includes marital status; highest education level completed; socioeconomic factors (household income, social class); occupation; generational cohort (e.g., Baby Boomer, Generation X); and number of children. Sometimes, demographic profiling incorporates geography; zip codes are commonly used to segment regionally.

Psychographics
   Psychographics classify individuals by interests, activities, and opinions (often referred to by marketers as *IAO variables*). These factors can include personality types, attitudes, hobbies, or lifestyle. Think about stereotypical high school cliques. When you hear "jock," "nerd," or "punk," you can probably envision a member of that group

and have a sense of that person's likelihood of buying a particular article of clothing or album.

Marketers often use a combination of psychographics and geographically bounded demographics to fine-tune their outreach strategies. PRIZM (*http://bit.ly/nielsen_segmentation*), a segmentation tool developed by consumer information company Nielsen (*http://www.nielsen.com*), sorts the population of the US into 66 personas ("Money and Brains," "Bohemian Mix," "Middleburg Managers") related to demographic and psychographic traits. Searching by zip code reveals the breakdown of various personas in a given area. While tools such as PRIZM are geared toward big brands focused on highly specific targeting, looking over the personas can help new entrepreneurs develop a sense of the factors that experienced marketers consider in their own segmentation studies.

BEHAVIORAL SEGMENTATION

Behavioral segmentation is another framework for identifying potential customers. If you're working within a crowded space (say, fitness trackers) and have chosen to compete by making a product with better battery life or greater ease of use, your most relevant market segment is likely not identifiable simply by gender or household income, but by technological sophistication. You might want to target frequent buyers, bulk buyers, or deal hunters. You might want to approach benefit or behavioral segmentation through exercises such as *persona building* around a given usage scenario: envision the most likely use cases for the product you want to produce, and then construct a persona that has the type of needs the use case fulfills.

A subset of behavioral segmentation is segmenting by buyer motivation. This involves understanding the type of benefit the user is seeking from the product. An individual who is considering buying a sweater, for example, might prioritize warmth, comfort, durability, price, or style. One or two primary motivators drive most purchasing decisions. Understanding the ones most common to your sector will help you strategically differentiate.

Within the B2B world, identifying your target customers involves different types of segmentation, but the concept remains the same. When identifying potential client companies, you might want to think about grouping customers by industry, standard industrial classification (SIC) category, size, market cap, or geography. On an individual level, you'll want to consider the needs and motivations of both the person who makes the purchasing decisions and the likely end user.

Usually, the appeal to the purchaser is based on economics, and the needs of the end users are technological or functional. It's possible to evaluate B2B customers using behavioral analyses as well. For example, some companies will buy at particular times of the year.

Chapter 4 discusses branding in much more detail, and Chapter 10 returns to marketing issues. For now, the takeaway is to understand the importance of thinking about your customer before prototyping, so that you can build smarter. If you'd like to learn more about market research, segmentation, and other aspects of brand-development marketing, MarketingProfs (*http://www.marketingprofs.com*) is an excellent source of information and tutorials.

Customer Development

So now you have a sense of how to segment your market and identify potential customer affinity groups. However, when building a company, you must do more than think theoretical thoughts about who your customers are, particularly if you're working in a crowded space. You need to get out there and talk to them.

This is one of the core tenets of Lean Startup that's as applicable to hardware startups as it is to software. You want to be doing customer interviews before you have any kind of functional prototype. You might not be able to give these potential customers access to a series of iterative alpha releases for them to play with, but you can show them designs or models and discuss potential features. Ideally, you can map each feature in your MVP back to a clearly articulated customer need. (See "Lumo BodyTech: A Case Study" on page 47, for an example of this process in action)

Hardware product development mistakes are particularly costly, in terms of both time and money. The goal of customer-driven development is to incorporate user feedback into the product development cycle in order to avoid building something that doesn't meet the needs of potential buyers. A good place to start the process is to reach out to individuals in several of the market segments you're considering and sit down with them to learn about their pain points. That will help you identify the real needs of your prospective customers—emotional, cognitive, and physical—and to discover how they're currently meeting those needs.

Here are some common questions for customer discovery interviews:

- Do you have Problem X?
- How much does Problem X cost you (in terms of time and money)?
- How are you currently solving Problem X?
- What do you like best about this existing solution to Problem X? What do you like least?
- What is the perfect solution to Problem X?
- Would My New Product solve Problem X?
- Why?

The last question—"Why?"—can turn a conversation from a surface-level survey into a discussion that unearths deeper needs, issues, and motivations.

It's important to phrase the questions you ask as neutrally as possible. You don't want to allow biased or leading questions to prompt your interviewees to respond in a certain way. It can also be difficult for entrepreneurs to really listen, particularly if the interview turns out to challenge their assumptions about a particular market or problem. Try to keep an open mind throughout.

For both B2C and B2B products, it's often ideal to sit back and observe as your target customers use whatever their current solution is to the problem you're trying to solve. You might learn more by watching their actual use case (and possibly hearing them narrate their issues as they go) than by asking them to articulate their needs. Your goal is to gain insights, even if they disprove your hypothesis.

Lumo BodyTech: A Case Study

Monisha Perkash is the CEO and cofounder of wearables company Lumo BodyTech. Its products, the Lumo Lift and Lumo Back, use sensors and software to help wearers monitor their posture and be more active. Here, Monisha details Lumo's approach to gathering early customer feedback as part of an iterative prototyping process.

"My cofounders and I came together without knowing what company we wanted to start," Monisha says. They were motivated by the desire to solve an important problem and make a positive impact in the world. The team spent six months exploring ideas, occasionally creating low-resolution prototypes for designs that were particularly appealing.

Monisha and her cofounders evaluated ideas based on three criteria: feasibility, viability, and desirability. Feasibility was an investigation into technology. Did the technology necessary to build the product exist, or was it almost there? Viability pertained to the likelihood that a product could become a self-sustaining business. Desirability focused on the needs of a theoretical customer. Would people want the device? Did it solve a pain point that customers would be willing to spend money or time on?

During this exploratory period, cofounder Andrew Chang started taking classes to help with his back pain. He found the experience life-changing and began to think about ways to incorporate technology into improving posture. The team had noted the increasing prevalence of sensors in their daily lives and were intrigued by the idea of helping people improve their health via a wearable sensor device.

The Lumo team began their process for evaluating the feasibility, viability, and desirability of the posture-device idea. They came up with the idea of a sensor that, when placed on the wearer's lower back, would help promote proper alignment. Using a post to Monisha's alma mater's alumni mailing list, the team recruited test subjects for early feedback. They made sure to select a diverse mix of genders and ages.

When gathering user feedback, the team avoided yes-or-no questions. They kept all questions broad and open-ended. "Tell me about how you're managing your back pain today." "What solutions are you currently using?" "How are they working out for you?" "What would you change about this current solution?" Above all, Monisha emphasizes the importance of asking *why*. She says, "You really want to discover the emotional need that person has around that issue."

The team continually refined their idea based on what they were learning about the customer pain point. Their lowest-resolution prototypes were made of paper, which they used to solicit feedback from 20 people. Several individuals suggested incorporating the sensor into an adhesive, similar to a nicotine or birth-control patch. The Lumo team incorporated that feedback into their next prototype, putting the sensor onto a sticker. It was a failure. "People conceptually thought they would want a sticker on their back," says Monisha. "In practice, they really didn't. It got icky."

The sticker design for the Lumo Back was gooey, collected lint, and was unpleasant for users with body hair. But most important, the team realized that it had a clinical feel. Since they wanted a more polished consumer experience, they recognized the need to rework the design. "We didn't want a 'clinical' product," Monisha says. "We felt that a strong consumer product was the best way to attract an audience, get feedback, iterate, and eventually put out new products."

After determining that the patch wouldn't work, they switched to a clip-on device that attached to the waistband of a garment worn by just about everyone: underwear. The closer the sensor was to the body, the more accurately it would read the tilt and angle of the wearer's back. However, feedback from the test group was clear: it was not a good user experience. Some commented that it was socially awkward. Others expressed concerns that it would fall into the toilet.

Eventually, the Lumo team decided to try a design inspired by a heart rate monitor. They bought heart rate monitors and wore them around their own waists for a week. After deciding that the monitor bands were comfortable, they once again reached out to their testers for feedback. Finding it positive, they decided to launch with the band form factor.

Despite conventional wisdom, Monisha believes that it is possible to do lean iteration as a hardware startup. "We went through 30 different design iterations and low-resolution prototypes before bringing this one to market," she says. She emphasizes the importance of being "scrappy and resourceful."

In addition to paper and other craft supplies, the team used readily available, inexpensive items to test designs. At one point, they went to The Container Store and bought tiny containers approximately the same size as the case that would hold their sensor. They attached the containers to elastic bands and wore them.

Once tester feedback was positive on both design and function, Monisha and her cofounders went to a prototyping facility and built 50 functional prototypes. Soon after, they launched a successful Kickstarter campaign for the LUMOback (now the Lumo Back) v. 1.

The Lumo BodyTech founders worked through dozens of ideas before finding one that both resonated with their desire to build something meaningful and filled a real need in the market. And after finding the right idea, they spent an extended amount of time developing the right form factor and feature set. Lumo has continued to listen to its target market as it has expanded its product line. The new Lumo Lift incorporates both features (upper-body posture feedback) and design (a more fashionable, customizable device) gleaned from Lumo Back customer feedback.

> While it is undoubtedly more difficult to apply Lean Startup principles to a hardware startup, with the right scrappy, resourceful approach, it can be done.

First-time entrepreneurs often wonder how to reach people in segments within which they don't have extensive contacts. An ideal way to do that is to use your existing network. It's worth building a LinkedIn presence to see who your second-degree connections are. A warm introduction—having a mutual colleague or acquaintance introduce you to the person you want to approach—is most likely to convince people to sit down with you. Beyond your friends and family, consider reaching out through an alumni network. Online, sites like LinkedIn, Facebook, and Meetup bring groups of like-minded people together. Blogger networks might also be helpful. Craigslist, Reddit, and Mechanical Turk users are often willing to take a survey (though typically for some type of compensation). Offline, reach out to local clubs or professional associations.

You can also leverage online marketing tools for product hypothesis validation on a larger scale: you can test and validate product features and price elasticity without building a single prototype. Make a few specialized landing pages, each offering your product, but highlighting a different feature set, design, or price point. That way you can identify the market(s) most worth pursuing in the early stage of your company, when you likely don't have extensive dedicated resources for marketing or outreach.

Buy Google AdWords to drive traffic to each landing page, targeting the customer segments you've identified as having the potential to be early beachhead markets. Using Google Analytics (or another analytics dashboard), track which keywords and sources bring people to your page. Observe the number of people who land on the page versus the number who provide you with an email address or other indication of interest in preordering. Services such as LaunchRock and Beta List, which enable users to sign up and "share" your company among their networks (in exchange for early admission to a beta), can also help you gauge interest and get the word out.

You should consider reaching out to people who have signed up with a more thorough survey about what interested them in your product. SurveyMonkey is an excellent tool for such a project. At a minimum, you are beginning to gather a list of interested customers to contact when announcing a crowdfunding campaign or taking preorders in the future.

If you're building a product with integrated software, consider using the software to validate your assumptions about the functions your users care about. Produce mockups, or even a rough alpha version of the software, and see what resonates.

To the greatest extent possible, you want to validate your assumptions about market size, desired product features, and price points well in advance of building your hardware MVP. Once you've started down the road to manufacturing, feature changes become increasingly difficult. Your goal is to be the best at your core use case from day one. You can incorporate the nice-to-have features in a subsequent release.

For more techniques, consider reading some of the many excellent resources that focus specifically on customer discovery and product development. Steve Blank's book *The Four Steps to the Epiphany* (K&S Ranch) is one of the classics. His blog (*http://steveblank.com*) is also rich with case studies (*http://bit.ly/blank_customer_dev*) and worksheets (*http://bit.ly/cust_discovery_skills*) that can help you conduct efficient interviews. *Lean Analytics* (O'Reilly), by Alistair Croll and Benjamin Yoskovitz, will help you use data to find your customers and develop the right product for their needs.

CHAPTER 4

Branding

As you go through the customer development and market research processes, you are collecting valuable information that will help you formulate a brand identity. Your brand is the personification of your company, and developing a strong brand is absolutely critical to your success. It builds a foundation for a long-term relationship with customers.

Branding can feel like a rather elusive concept, as the value-add of a strong brand is difficult to quantify and measure. Startups often undervalue the importance of building a brand, particularly if the founding team is strong on tech but has no marketing or sales experience (as is often the case). In any early-stage company, there is much to do and precious few resources to do it with. There is never enough time, money, or people, so most founders put all of their resources toward nailing the product. Branding, marketing, sales strategy…those are problems to push off until a later date.

Don't make this mistake.

The problem with this approach for a hardware company is that your product will be competing for shelf space (digital or physical) with established players. When you are on a physical shelf, there is no website with help text or comparison charts that can explain the virtues of your product. Your package messaging must be appealing enough to convince a busy shopper to put your widget into her cart. If your product is on the shelf next to one of similar price manufactured by a competitor who has better name recognition, your product is at a disadvantage. People have many choices, but little time. They're going to grab the product they've heard of, or the brand they are loyal to.

While startups often neglect brand building, Fortune 500 companies prioritize it. They treat branding as a critical facet of their business strategy. A study by Interbrand and JP Morgan determined that, on average, *brand* accounts for close to a third of shareholder value. Though valuing intangibles is notoriously difficult, the study says:

> *The brand is a special intangible that in many businesses is the most important asset. This is because of the economic impact that brands have. They influence the choices of customers, employees, investors and government authorities. In a world of abundant choices, such influence is crucial for commercial success.*

In 2013, Apple was the most valuable brand name in the world, worth an estimated $98.3 billion (*http://bit.ly/brand_values*). Google was second on the list, at $93.3 billion, and Coca-Cola was third, with $79.2 billion.

Brand equity is the monetary value that comes from having a recognizable brand. Marketing experts have found that positive name association enables a company to justify a price premium over similar goods. Think of the product options on the shelf of your drugstore: is Clorox bleach better than store-brand bleach? Is Advil better than generic ibuprofen? In both cases, the generic is exactly the same product, but it costs more. And yet, people still buy it. That price premium is Clorox's and Advil's brand equity.

In the world of devices, your product is not exactly the same as your competitor's product. At a minimum, you likely have a few different features and a different design. But when customers are deciding what to buy, they are purchasing a product to meet a need or fulfill a want. If either your device or your competitor's device will satisfy their objectives for approximately the same price, brand becomes a powerful differentiator.

Emotional responses to products matter. According to Alina Wheeler's seminal brand-strategy guide *Designing Brand Identity* (Wiley), brands serve three primary functions:

Navigation
A strong brand helps customers make a choice when presented with a wide array of options.

Reassurance
In a world with so many choices, a brand reassures customers that the product they've chosen is high quality and trustworthy.

Engagement

Brand visuals and communications make customers feel that the brand understands them. The result is that customers identify with the brand.

A recognizable brand can help a company increase (or defend) its market share by inspiring trust and enhancing the perception of quality. Having customers who identify with your brand engenders loyalty.

Brand loyalty is important in an industry in which product turnover is high. Within a few years, new technology renders many hardware products outdated. People upgrade consumer electronics (phones, music players, cameras, speakers) every few years. High-quality software products are developed to engender what's known as *lock-in*: over time, customers become accustomed to the feature set, learn advanced shortcuts, store their data, or create large libraries of files. They become loyal customers and will often pay to upgrade the software as new versions are released. They have invested time and energy in learning how to use it, and this makes them reluctant to switch to a new product. This phenomenon is particularly prevalent in enterprise software, where corporations want to ensure continuous availability of data and internal documents. They don't want to risk losing document integrity porting to a new product.

With most hardware products, this kind of lock-in is difficult to achieve. Unlike with software, where you can push an update that incorporates new technology (and suggest that your users buy it), hardware upgrades eventually require new physical components encased in new plastic. At the point of purchase, whether on the shelf or online, your customers will be confronted by alternatives. Strong branding generates loyalty and makes customers more likely to have you top-of-mind when they intend to purchase; even when they're *not* actively intending to purchase, you want your brand to be top-of-mind.

Besides loyalty, a strong brand gives you leverage when expanding your offering into a new category. This is a particularly important consideration for connected-device startups or for wearables companies that aspire to be platforms. According to Sean Murphy of product development consultancy Smart Design (see his discussion of branding and design in "From Conception to Prototype with Smart Design: A Case Study" on page 56), "There really aren't 'connected products'; there are just connected brands. You don't just experience the product; you experience an ecosystem."

From Conception to Prototype with Smart Design: A Case Study

Sean Murphy was previously the Director of Design Engineering at Smart Design, a design innovation consultancy with expertise in both digital and physical products. Its client roster includes powerhouse companies that produce both designed products (Oxo) and consumer electronics (Flip, Toshiba). Here, he tells us about the process Smart Design uses to take its clients from conception to prototype.

The Smart Design team began the process with a qualitative user research study. "We start with a hypothesis and a design instinct around a product, and we put that in front of who we think the target audience is," Sean says. Participants in the qualitative study often come through recruiters, to ensure an unbiased and independent set of opinions. There are typically 8 to 12 participants, representing a range of target customer personas. Occasionally, if a client has a specialty focus (e.g., for a product geared toward the deaf population), the client will provide the participants. The team will set up in a potential user's home and watch him go through the usage scenarios. The product is often in very raw form, sometimes just a paper prototype or illustration.

The emphasis in these early conversations is on identifying the important functional elements that the product should have, and how the product will fit into the user's daily life. "This builds a narrative for the design vision," Sean says. The team's goal is to assign a relative value to the different functional elements according to how important they are for solving the customer's problem. The resulting framework is the basis for the design. At this point, it's still flexible, but it's a reference for what users care about and why.

The design research team might present brand elements to the participants during the qualitative study, addressing how a product makes them feel or what it means to them. The team wants to understand what is important, and why. "We try to make it abstract, to allow them to pour their own narrative onto the product," Sean says. "You want to hear things like, 'This is like this other product that I had 10 years ago and loved for these reasons.'" The team might also ask the participants to rank competitive brands, or do some association exercises. This helps the client gain insights into tone and positioning.

The Smart Design team uses the information gleaned from the qualitative study to create one or more design directions for the product. If the client's schedule and budget allow for it, the Smart Design team will return to the qualitative study participants (and bring in some new ones) to show them several possible prototypes and solicit more feedback. The goal is to observe user reactions. It's a learning and refinement process, geared toward guiding further development.

The team presents multiple versions of a product to ensure that the user doesn't think of a prototype as *the* product. "Perhaps you have a strong product ID, but you're taking refined prototypes to the field that are more about usability than ID," Sean says. "From the design perspective, you go to the field to learn and refine rather than validate." This second-round conversation can also help develop a greater understanding of brand perception.

At various points in the design process, the client might express a desire for a more rigorous quantitative study with a larger sample size. The quantitative analysis is often a web-based survey with questions such as, "On a scale of 1 to 10, how important are these three features?" and "Would you rather have feature X or feature Y?" According to Sean, "It's designed to be an A/B test for a physical product."

A quantitative study can give the team additional data to help them anchor design elements and reinforce what was learned from the qualitative interviews. "One of the big challenges for design is to make decisions not feel so subjective," Sean says. "There are always elements of subjectivity in a design process, but when you're trying to make a decision as to something that influences the cost, it's good to have a more rigorous rationale."

After the design principles are established, product development continues along the industrial design process, blending together features and interactions with distinct styling directions. In general, it takes one to three months to get to a design prototype, depending on the complexity. Wearables and products that are ergonomic can require multiple iterations and user studies. Often, a company has not yet engaged a contract manufacturer, so the design is subject to revision as the internals of the product change.

Many products have technology requirements that act as a limiting factor on the design, so it's important to work with a designer who understands the technological limitations up front. A contract manufacturer will do its best to work with the design files it's given, but designers without experience in a particular market might neglect to incorporate certain elements—say, holes for venting heat. The manufacturer will add in the necessary holes, possibly at the expense of aesthetics. The smaller the product, the greater the possibility that a design will be unable to accommodate the necessary internals. Wearables are particularly challenging.

Once a product gets to tooling, it's difficult and costly to make changes. To mitigate this type of problem, Smart Design has a technical team that understands both frontend and backend needs. Sean says, "One of the advantages to working with someone who stays engaged all the way through the process is that the same designer is there when something changes, which almost certainly will happen."

Working with a consultancy for the full design-to-production process can cost well into the six-figure range, which is often prohibitive for startups. Since the contract is typically priced by phase (in which each phase is tied to a deliverable), a startup might want to approach a firm with its qualitative and quantitative research already done. One of the value-adds of working with a firm that has technological, design, and branding experience all under the same roof is that simultaneous development of all three will help the startup ensure that no one aspect is treated as an afterthought.

Jawbone started out as company called Aliph; Jawbone was the name of its first wireless headset. It released several subsequent models of headsets over the years—Jawbone Prime, Jawbone Icon, Jawbone Era—before eventually dropping the name Aliph and rebranding as simply Jawbone. Under the successful Jawbone brand, the company expanded from headsets into portable speakers (Jawbone Jambox) and then a fitness device (Jawbone Up). The packaging for each of these devices reads: "ERA by Jawbone," "UP by Jawbone," etc.

Prioritizing brand recognition gives a company a leg up on a new launch, in terms of both awareness and perception of quality. Regardless of the distribution channels you pursue, you won't sell many widgets if you're an unknown quantity.

So how do you build a recognizable brand?

Your Mission

First and foremost, a brand must have a *mission*. Earlier in this chapter, we touched on the importance of identifying a problem that truly motivates you and resonates with others. Your mission is what your company is doing, why, and for whom. You should be able to articulate this succinctly in the form of an *elevator pitch* or *mission statement* (sometimes called a *mantra*). Your brand mantra is a statement of why you exist. Consider the following examples:

Apple
"Committed to bringing the best personal computing experience to students, educators, creative professionals, and consumers around the world through its innovative hardware, software, and Internet offerings."

Microsoft
"To enable people and businesses throughout the world to realize their full potential."

Nike
"To bring inspiration and innovation to every athlete* in the world. (*if you have a body, you are an athlete)"

Google
"To organize the world's information and make it universally accessible and useful."

These sentences distill each company's intent and purpose into a single statement that represents the core of the brand's identity. All other facets of branding—personality, assets, and experience—are outgrowths of this statement of purpose.

The company's products also reflect this purpose. Consider Apple's and Microsoft's mission statements in the context of their product lines. Apple builds beautiful products and prioritizes a seamless user experience. Microsoft builds exemplary productivity tools used in enterprise companies all over the world.

Knowing your mission helps your company in both an internal and a public-facing capacity. Internally, it serves as a guide for employees to know what they stand for and what they're working toward. It provides a company with a framework to evaluate strategies and products: to what extent does a specific action or product release advance your mission and align with your core values?

Sean Murphy (see his discussion of branding and design in "From Conception to Prototype with Smart Design: A Case Study" on page 56) points out:

> There's a tremendous pressure to cut corners when you're a startup. Having strong brand principles gives you something to refer back to during development. As the product evolves, you're going to make decisions with respect to what you understand your brand to be. The software, the service

> components...it will all be viewed in the context of answering the question "Who are we?"

Publicly, your mission statement is a communication tool that frames your brand in the minds of consumers. You are telling people what you stand for and why you exist. In his book *Grow* (Crown Business), marketing expert and former Procter & Gamble global marketing officer Jim Stengel advocates that a company should have a *brand ideal*, a "higher-order benefit it brings to the world" that satisfies a fundamental human value that improves people's lives. In his own words, people's lives can be improved by engaging five fundamental human values:

Eliciting Joy
Activating experiences of happiness, wonder, and limitless possibility.

Enabling Connection
Enhancing the ability of people to connect with each other and the world in meaningful ways.

Inspiring Exploration
Helping people explore new horizons and new experiences.

Evoking Pride
Giving people increased confidence, strength, security, and vitality.

Impacting Society
Affecting society broadly, including by challenging the status quo and redefining categories.

These ideals elicit emotional responses. Emotional connections lead to deeper relationships with customers. Brand communication skills expert Carmine Gallo has interviewed numerous CEOs about what their brands stand for (http://bit.ly/brand_meaning). He recalls Tony Hsieh, founder of Zappos, replying with one word: "happiness." Richard Branson, CEO of Virgin Group: "fun."

New founders might be skeptical of the real impact of values and ideals on important metrics, such as sales numbers. While a direct link is difficult to quantify, Stengel cites a study that examined the connection between financial performance and customer engagement and loyalty over a 10-year period. The researchers looked at 50,000 brands and found that in the minds of consumers, the 50 top high-growth brands were linked with an ideal. These 50 companies (called the "Stengel 50") grew three times as fast as their competitors over the 10-year period. One example of such a company is Pampers (*http://www.jimstengel.com/wp-content/uploads/2013/11/GrowChapterOne.pdf*):

> *Pampers' brand ideal, for example, its true reason for being, is not selling the most disposable diapers in the world. Pampers exists to help mothers care for their babies' and toddlers' healthy, happy development. In looking beyond transactions, an ideal opens up endless possibilities, including endless possibilities for growth and profit.*

You can't be all things to all people. A well-defined set of values and authentic messaging will help you attract customers who share your values and vision, and care about what you are trying to do for the world. As a startup, it can also help you find investors and potential employees who are aligned with your mission.

Brand Identity and Personality

Putting in the effort to arrive at a deep understanding of your brand's values and mission will help you define and project a coherent brand identity. As defined in branding expert and author of *The Brand Gap* (New Riders) Marty Neumeier's The Dictionary of Brand (*http://bit.ly/dict_of_brand*) (an excellent resource for brand builders), brand identity is "the outward expression of a brand, including its trademark, name, communications, and visual appearance."

Brand identity is the sum of all of the parts. Identity is deliberately constructed by the company, with the goal of ensuring that customers both recognize the brand as an entity and can articulate how it differs from the competition. *Brand image* is the consumer's perception of this identity: how the market views your brand. You want the market view to align with the impression you are trying to create.

As Neumeier puts it (*http://bit.ly/the_brand_gap*), "[A brand] is not what **you** say it is, it's what **they** say it is." To have an impact on your brand image, you must actively manage your brand identity.

Professional brand strategists use a variety of methods and frameworks to define and shape identity. In *Designing Brand Identity*, Alina Wheeler breaks down the process into five steps:

Conducting Research
   Fully investigate the existing perception of the brand, both in the market and in the minds of *stakeholders* (constituents who have a vested interest in a company, such as employees, investors, customers, partners, etc.).

Clarifying Strategy
   Define goals, identify key messages, and determine appropriate strategies for naming, branding, and positioning.

Designing Identity
   Define a unifying "big idea" and develop a visual strategy.

Creating Touchpoints
   Produce visual elements, refine the look and feel, and protect trademarks.

Managing Assets
   Develop and implement a launch strategy to unveil brand elements, define brand standards, and establish guidelines to ensure consistency.

This process might seem like an extensive undertaking, particularly for a resource-constrained startup, and it can be a long and time-consuming process. But even in the days before a startup has stakeholders aside from the team, it should be thinking about brand identity. Jinal Shah of marketing communications firm J. Walter Thompson (JWT) discusses ways that startups can streamline this process in "Brand Building for Startups: A Case Study" on page 62.

Brand Building for Startups: A Case Study

Jinal Shah, Global Digital Strategy Director at J. Walter Thompson, has 10 years of experience as a brand and digital planner. Throughout her career she has worked with both large brands and startups. In this case study, she shares the highlights of the branding process and how startups can do this most effectively with limited resources.

Typically a company writes a *brief* before engaging a branding agency. The brief captures the objective of the assignment, the desired outcome, any competitive insights, and mandatories that will help the branding process. "This is a foundation document. It must also convey key information about your company. The story of how it was started, etc."

The agency will then begin a stakeholder audit and interview at length the founders, the employees, and often also the customers/users (both the current customers and those the company aspires to attract). The goal of this audit is to get a sense of the values shared across all groups—because they are the basis for the underlying values of the brand. This audit will also uncover any discrepancies in the brief and other weaknesses and consistency issues—which all contribute toward building a strong brand and hence are necessary to identify and address early on.

Jinal says, "Consistency is absolutely key when articulating and designing your brand. It is important that all the stakeholders are aligned."

One of the things this initial research often produces is a clear understanding of values: a document that says "who we are; who we are not." A company might say something along the lines of "we are global, but we are not corporate" to indicate its size and outlook but also refer to its informal and approachable personality.

"You'll start to see patterns emerging...words and phrases that appear repeatedly," Jinal adds. Those attributes are the foundation of any company's corporate culture. Once the culture is identified and defined, the brand's personality, assets, and messaging can be designed as manifestations of those ideals.

Once the brand values are defined, the next step is to address the brand positioning. "Brand positioning is about identifying the space you want your brand to occupy in your audience's minds," Jinal says. "What do you exist in the world to do? You have to nail your answer to that. Everything cascades from that."

At this point, JWT often engages in extensive competitive and landscape analysis to ensure that the positioning it defines for a client is unique, well differentiated, and future-proof.

Positioning is about the brand, not the product that the company offers. "Focusing positioning on the product commoditizes your brand and is short-sighted," Jinal says. It will box you into a category, making it difficult to elevate and leverage brand recognition if you expand into a new market or pivot for the future. She continues, "The goal of brand building is to attach emotional value to the brand. When there is a competitor, a choice in the market, the emotional connection is what will make your customers pick you over anyone else."

This stakeholder audit and research takes a few weeks and can cost from $30,000 to $50,000 with a reputable firm. Most early-stage startups don't have the resources to pay for this formal branding process. "Startups have fewer resources but more agility," Jinal says. "What takes a large company months, a startup can often accomplish in a few weeks."

For starters, there are fewer stakeholders to interview. And in general, startups exist because founders have already identified a need (a "white space") in the market that they are working to solve.

A startup with limited funds can work through an effective brand-building process without using an agency. To derive brand values and personality, the founding team should start with the kind of people they are. "Start with cultural attributes that are very true to you," Jinal says.

Authenticity is important. In the early days, the founders (the visionaries) are the most effective mouthpiece for the brand, especially online. They will not be effective if they're trying to be something they're not.

"The thing to remember about brand building is that everything communicates—so once you know who you are and what your values are, infuse them through everything you do and produce," says Jinal. Your brand values should dictate how you onboard new employees and handle customer complaints to the color of the paint on your walls and the brand of coffee in your office kitchen.

A community manager is a key hire in bringing your brand to life externally. The founders may define the brand, but the community manager will be the person who most consistently has the opportunity to convey it to users. If you don't have a community manager or head of marketing yet, make sure someone on the team is an excellent communicator. That person should be responsible for infusing the brand's personality into touchpoints until a dedicated hire or agency partnership is in place.

Once your brand is defined, you'll need to build assets. For example, if you are a hardware startup, you may need packaging that shows off your product but also communicates your brand. When you are meeting with the packaging agency, bring packaging materials and examples of different types of packaging that resonate with you and reflect either your brand or brand values. According to Jinal, "If you're engaging a creative person you want to inspire and challenge them, but also educate them about your brand and your preferences to get the best work out of them."

Early brand building takes a lot of time and discipline. It should be a serious, process-driven endeavor. Startup founders put a lot of energy into product development and trying to attract customers. Identifying brand values and

> knowing who you are is equally important. Your ultimate goal is to sell your brand—the values, vision and emotional connection—not your product.

A fundamental component of brand identity is *brand personality*. Emotions are involved in the purchasing process, as the customer has a need or desire and is looking to fulfill it. Consumers buy products that fit the perception they have of themselves, or the way they wish to be perceived by others.

A successful brand has a clearly defined personality that appeals to or resonates with its target customers. In *The Dictionary of Brand*, Marty Neumeier defines brand personality as "the character of a brand as defined in anthropomorphic terms." Brands can be kind, funny, masculine, elegant...the possibilities are endless.

Agencies use several common frameworks to help their clients identify a brand personality, some of which a research-constrained startup can carry out independently. One is an *archetype study*. Archetypes are a universal model for a personality, such as "the Joker" or "the Rebel." Hardware startup Contour Cameras took this approach, which founder Marc Barros described in a blog post about the process (*http://bit.ly/building_a_brand*): "Defining a brand is like defining a person. No different from how you would describe a friend, brand attributes are the adjectives you choose to define the personality of your brand."

Using the 12 archetypes defined in Margaret Mark and Carol Pearson's book *The Hero and the Outlaw* (McGraw-Hill), the Contour team identified the persona that fit their desire to facilitate imaginative self-expression: the Creator. Using the Creator as an anchor, they focused their brand on creativity. They worked on producing a product that would be loved by Creator types: artists, innovators, and dreamers.

In psychological research, the five-factor model (*http://bit.ly/big_five_traits*) asserts that there are five basic dimensions of human personality (called "the Big Five personality traits"): openness, conscientiousness, extraversion, agreeableness, and neuroticism. In a similar vein, Stanford marketing professor Dr. Jennifer Aaker believes that brand personality can be broken down into five core dimensions (*http://bit.ly/brand_personality_dims*): sincerity, excitement, competence, sophistication, and ruggedness. Within each of these dimensions are facets and traits that increase the specificity of the description, as shown in

Figure 4-1. While some academics have criticized Aaker's scale as biased toward American culture, these lists of traits are a solid jumping-off point for companies looking to define their brand personality.

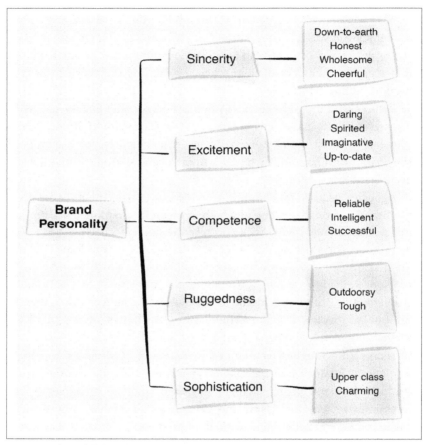

FIGURE 4-1. Aaker's brand dimensions

Another approach to the process involves listing adjectives that describe your target customer. Some branding agencies have their clients work through association exercises, such as "If you were a car, what kind of car would you be?" or "If you were an animal, what kind of animal would you be?" Or you could try personification: imagine your brand as a real person and visualize what she would look like and how she would act.

Products and messaging keep changing over time as a company evolves, but brand values and personality generally remain the same. Jinal Shah says:

> You want to have a little bit of room to evolve and have fluidity in the market, but there is still a core set of values that must remain consistent. Think about you, the person. There are certain things that you would do or say, and certain things that you would not do or say. People who know you can probably distinguish between the two. That's how your personal brand gets built. It's no different for a company.

Brand Assets and Touchpoints

Brand personality is brought to life via *brand assets*. Assets include the brand name, logo, tagline, graphics, color palettes, and sounds...sometimes even scents and tastes.

Naming your company is incredibly important. A name is the most frequently used brand asset. Customers see it and hear it. They speak it when they tell their friends about you. Many blog posts have been written about naming software startups: common suggestions include keeping the name short, choosing something with an available domain (or, at worst, a unique modified domain with the name in the title, such as *GETstartup.com* or *startupHQ.com*) and social media handles, and selecting something that is easily pronounceable and searchable (don't remove too many vowels!). A name should be distinctive and memorable. It must also be legally available and not trademarked by anyone else.

One important difference between naming a hardware startup (or product) and a software startup is that the name is more likely to appear on packaging, and possibly on the device. Clarity and readability are critical. Can you tell from a glance at the name what the product does? The ability to extend a product line in the future is also something to consider. Apple has done this exceptionally well. The "iDevice" naming convention has taken it through decades of hardware products (iMac, iBook, iPod, iPhone, iPad), and links the hardware to the software (iTunes).

Some companies use their founder's names. That way a corporate brand can benefit from being associated with a charismatic founder's personal brand (Beats by Dre), but there is a risk of negative associations if the founder endures any personal scandal or hardship (Martha Stewart). Also, be wary of choosing a name that is a generic word. While many successful companies have such names (such as Square and Nest), it takes hard work to associate the word with your company in the minds of consumers. You'll also spend quite a bit of money acquiring the domain.

Begin your naming process by deciding what you most want the name to evoke. You can use the work you did when identifying your values and personality. Choose several words or phrases that capture the brand essence you're going for, and begin to brainstorm. There are many categories from which to pick a name:

- Emotions that you want your product to evoke
- Locations where people are likely to use it
- A distinctive physical characteristic of the product
- A metaphor that represents your user or your product
- A verb related to your product's functionality

It's helpful to work through this process with a group of people. Word association and building off of the creativity of others can make it a lot easier and more enjoyable.

Once you've come up with several names that you like (and checked them for potential trademark infringement), it's time to test them. San Francisco naming consultancy Eat My Words has a series of criteria that it calls the SMILE & SCRATCH test (*http://bit.ly/smile_n_scratch*).

The SMILE test checks to see if your name has the following important qualities:

Suggestive — evokes a positive brand experience
Meaningful — your customers "get it"
Imagery — visually evocative to aid in memory
Legs — lends itself to a theme for extended mileage
Emotional — resonates with your audience

And the SCRATCH test helps you determine if it should be scrapped:

Spelling-challenged — looks like a typo
Copycat — similar to competitor's name
Restrictive — limits future growth
Annoying — forced
Tame — flat, uninspired
Curse of knowledge — only insiders get it
Hard-to-pronounce — not obvious, unapproachable

If your name passes these two tests, it's time to consider testing it in the wild. Some entrepreneurs choose not to, instead just trusting their own judgment. Others solicit feedback solely from friends, family, or their (early) community. Naming and branding experts advise field testing, because it's often difficult for insiders to recognize that their name has a "curse of knowledge" problem or is inscrutable to a general audience. They test names with focus groups. Common questions that naming experts ask include:

- What do you think this business does?
- Can you spell it? (Can you pronounce it?)
- Does this name remind you of any particular product?
- What do you think of when you read the name?

If you're bootstrapping this process and decide you want to get consumer feedback on your product name, online surveys are a less expensive and less time-consuming way to go. You can reach a wide audience for a minimal fee using a service such as Mechanical Turk (*https://www.mturk.com/mturk/welcome*) or Crowdflower (*http://www.crowdflower.com/*) to find survey participants. Some founders try A/B testing names in AdWords (*http://adwords.google.com/*), driving traffic to landing pages.

Still, there's no substitute for in-person feedback whenever you can get it. Marc Barros, the founder of wearable-camera maker Contour, describes its process (*http://bit.ly/picking_a_name*) of using the "Bar Test": say your company name to someone in a noisy bar. See if they understand what you do. If they can't pronounce or spell the name, consider it a failure. (For more lessons learned from Marc's experience with company naming and branding, see "Naming Contour and Moment: A Case Study" on page 69.)

Naming Contour and Moment: A Case Study

Marc Barros is the founder of Contour cameras, which made a wearable/mountable point-of-view video camera until it shut its doors in August 2013. He recently launched his new company, Moment, via a successful Kickstarter raise. He also organizes the Hardware Startup Workshop, a daylong series of talks by industry experts that helps founders learn best practices and avoid potential pitfalls. As a founder who has experienced both success and failure, Marc has a diverse perspective on the importance of early brand building for hardware startups. Here, Marc talks about the challenges the Contour team faced and how he's applying the lessons he learned toward building Moment.

At Contour, Marc says, "We got branding horribly wrong at the very start." Contour was the third name for the company. The first, chosen when the founders were both in college, was 20/20. They decided they didn't like that, and switched to VholdR. "We were trying to come up with a new name that worked for the camera, the software, and this new social community we wanted to make around video," Marc says. Unfortunately, people couldn't spell it or pronounce it—not even members of the team. They didn't have the money to hire a naming firm, choosing instead to put their funds into product design.

The absence of social networks in the early days of Contour's existence meant that their marketing strategy relied heavily on in-person events. They targeted the action sports and lifestyle communities. "We used a lot of event strategy," Mark says. "We would go to mountain biking events, skiing events, snowboarding events. We had an RV driving around, and reps that would go with us to help canvass the area."

Contour also built a network of athlete influencers, giving them product and sponsoring them. At various points in the company's evolution, the product was sold under the 20/20 and VholdR names. The camera achieved visibility within the target market, but the multiple rebrands meant that there wasn't much continuity, which is critical for building up a brand identity in the minds of consumers.

The name changes resulted in lost time and lost market share. A full five years into the existence of the company, the team got serious about branding. They had managed to gain some traction (and earn some revenue) despite the setbacks, and they began to work with an agency to shape the Contour brand archetype and story.

Unfortunately, competitor GoPro hadn't made the same missteps. Founded at approximately the same time, and with a similar offering geared toward a similar customer, GoPro continued to gain market share and move ahead of its smaller rival. After shutting down Contour, Marc wrote an excellent post-mortem on his blog (*http://bit.ly/contour_postmortem*) detailing the differences between GoPro's branding strategy and Contour's. In the post, he describes learning the lesson of "brand first, distribution second." Focusing on retail sapped Contour's resources, leaving little available to drive consumer demand. GoPro focused on emotionally connecting with customers, creating an aspirational brand identity through the use of thrilling action videos shot with the product. GoPro built a movement.

"Contour wasn't the best in the world at one thing," Marc reflects. "We were pretty good at product, we were okay at brand, okay at distribution, but we weren't unbelievable at one thing." The lack of a crystal-clear value proposition, combined with nebulous positioning, made competing with its well-branded rival an insurmountable challenge.

Moving on to Moment, Marc has applied the lessons learned from the Contour experience. This time around, "We started by understanding who we are and why we're doing this," Marc says. From the beginning, the team has focused on identifying their core values, making those values the foundation of the company they're building. They have worked at understanding the companies that have come before. According to Marc, "This brand is anchored in the history of lenses." The design elements, the name, and the concept are all part of a cohesive vision: segments of time. Moments.

"We all took photos, put them on an inspiration wall," Marc says. "We thought about what the photos made us feel and realized that photos were really about the moments." Photography is about remembering moments and sharing moments, and the Moment team wants to facilitate that emotional experience with easy-to-use, powerful lenses that attach to mobile phones. Early indications from the Kickstarter raise suggest that the vision resonates with many people. The team set a goal of $50,000 and raised $451,868.

Moment is focused on what Marc calls "picture takers." *Photographers* are professionals; *picture takers* are the people who have day jobs but love using their mobile phones to document their lifestyle, their weekend adventures, their kids. The success of Instagram suggests that there are quite a lot of them. "We don't really believe in building customer profiles: 'this is Jo and she works here and these are the products she buys.' We look for the intrinsic needs," Marc says. Picture takers are creatives who happen to snap photos in their spare time, so the Moment team is focusing on creating mobile lenses that prioritize speed and convenience, ideal for everyday use.

"It takes a couple of years to really know your brand—to get the details right, to know the customers, to understand why they're buying," Marc says. The team is diving into their Kickstarter backer profiles to learn more about their customers—how they heard about Moment, and why they bought it. They're using this data to better understand the demographic and psychographic characteristics of their target market.

However, they'd begun to reach out to potential early adopters and influencers long before attempting a crowdfunding campaign. "We found a group of picture takers local to us, so we could go have coffee, show them the product, and get their feedback," Marc says. That group has helped the team along the way. Marc adds, "They buy into the vision—it's so much more than 'here's a widget.' They're interested in who the company is and what we believe, so they're helping us along the way."

The Moment team has identified the one thing in the world that they will be the best at: their mission is to deliver the best mobile photography products. With that goal in mind, they're laser-focused on product development. "We're not working on retail distribution, because that would require hiring people and spending money and margin on it," Marc says. "We're focusing on building a team that's unbelievable at product."

While the Contour experience was painful, Marc has applied the lessons he learned to skillfully crafting a clear brand mission and identity for his second company. The Moment team has prioritized branding from day one, and they're off to a great start.

Your brand name and personality are the cornerstones for the rest of your visual assets: logo, graphics, color palettes, and icons—and you have many creative choices to make. Certain colors evoke specific emotions. A logo can be a word ("Google"), a picture (Apple's apple), or a combination of the two (the 1992–2011 Starbucks logo). The goal is to create representative visuals that are immediately recognizable and memorable and that retain their impact when displayed across different mediums.

The sensory experience should be coherent and in line with the brand personality. For example, if your brand personality is elegant and sophisticated, a low-resolution cartoon animal logo would seem incongruous. Visual assets should fit cohesively with the look and feel of your product, including any software or apps experienced by the end user. For a good example of precise execution around visual identity, check out Google's Visual Assets Guidelines (*http://bit.ly/google_visual_assets*).

Producing quality brand assets takes time and costs money. If you don't have the resources or desire to work with a branding agency, your designer is the person most likely to be responsible for this process. If you don't have someone with design experience on the team, hire an expert contractor. Brand assets might evolve over time, but you'll want a polished appearance from day one. And just as with everything else in life, you get what you pay for. Expect to pay $50 to $100 an hour for a professional.

The points at which brands interact with consumers are called *touchpoints*. To identify touchpoints, think about your outreach channels: media, packaging, advertising, environment (e.g., stores). Possible touchpoints include websites, emails, apps, blogs, TV, trade shows, exhibits, print materials, circulars, billboards, videos, kiosks, retail shelves, social media...basically, anywhere in the physical or online world that a consumer might encounter your brand.

Touchpoints can be quick or sustained, personalized or mass-market, real-time or static. Product design consultancy Hello Future has an excellent matrix that illustrates this *(http://brandtouchpointmatrix.com/)*, as shown in Figure 4-2. Note the position of "campaigns" (advertising) at the bottom left: quick and mass-market. Up at the top right are "platforms," which are long-term and sustained.

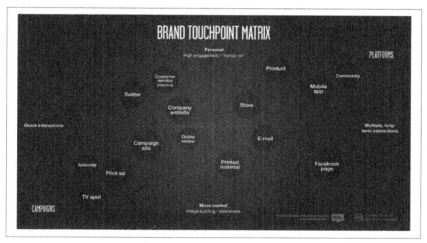

FIGURE 4-2. Hello Future's Brand Touchpoint Matrix (used under Creative Commons Attribution-ShareAlike license *(http://creativecommons.org/licenses/by-sa/2.5/)*)

Consumers might encounter touchpoints at the prepurchase, purchase, or postpurchase stage. The goal of a prepurchase interaction is to shape a consumer's perception of your brand and communicate your value proposition. You are attempting to increase the consumer's likelihood of buying your product. At the point of purchase, the goal is not only to make the sale, but also to establish a deeper relationship.

Postpurchase communications are geared toward building loyalty, ensuring satisfaction, and turning customers into brand evangelists. Across all touchpoints and at all phases of the purchasing decision, you're selling the brand, not just the product.

The consumer's experience with your company at any given touchpoint is known as the *brand experience*. The total of interactions at various touchpoints over time results in a cumulative experience. Both individual and cumulative experiences matter. Chapter 10 discusses metrics in more detail, but it's worth noting here that you will want to have mechanisms in place to gather data about your customer's experience at each touchpoint.

As soon as you begin to communicate with the public, you should be striving for a consistent presentation of your brand across all possible touchpoints. Consistency applies to both the message content and tone, anywhere that your brand name or logo appears.

Positioning and Differentiation

Your *brand position* is the space in a given market that you occupy in the minds of consumers. It's your unique niche. Customers mentally rank brands. How would you rank McDonald's, Burger King, and Wendy's? Understanding what is special about your product and what position it occupies in the market is a critical step toward gaining market share among your target audience. In the soft-drink market, 7UP was number three behind Coca-Cola and Pepsi. When it began to identify itself as the "UNCOLA," it made itself the leader in a different category: alternatives to cola.

Positioning is a function of three elements: customers, competitors, and a characteristic. The *customers* are the target market you're going after: whom are you trying to reach? The *competitors* are the other companies that are already in the market. The *characteristic* is your differentiator; in marketing, it's called the *point of difference*. Early in the process of market research, you identified ways to differentiate your product from the competition: by features, price, or some other factor. A powerful point of difference for positioning your brand is something that's both *defensible* (your competitors can't quickly replicate it) and important to your target customer. As mentioned above: you can't be all things to all people. Specificity is key.

Marketing expert Geoffrey Moore has a template for synthesizing a thorough positioning statement in his book *Crossing the Chasm* (HarperBusiness):

> *For (target customer) who (statement of the need or opportunity), the (product name) is a (product category) that (statement of key benefit— that is, compelling reason to buy). Unlike (primary competitive alternative), our product (statement of primary differentiation).*

This statement incorporates all of the different preproduct considerations we discussed in Chapter 2 into one sentence and synthesizes them into a single message. Chapter 10 returns to positioning and differentiation within the context of marketing. In the preproduct phase, an understanding of whom you are building for and what position you wish to occupy in the market will provide you with a framework against which to make product decisions.

Brand development is an ongoing process, and it's worth your time. If you create a brand synonymous with quality and a joyful experience, the resulting loyalty and satisfaction will translate to increased customer retention...and increased revenue. Recognizing the importance of branding to the overall success of the company, some startups make their first in-house hires early. For example, Nest's third employee was sales and marketing expert Erik Charlton (see how Nest built its brand right from the start in "Nest Branding: A Case Study" on page 75).

Other startups hire a branding agency. Regardless of the approach you choose, it's important to begin thinking about branding as early as possible. From the minute you exit stealth mode and the public becomes aware of your existence, consumers—as well as potential hires, partners, and investors!—are forming a perception of who you are. The more actively and thoughtfully you shape that perception, the greater your chance of becoming a successful company.

Nest Branding: A Case Study

Nest is a connected-devices company on a mission to reinvent "unloved," but important, products in the home. Its first two offerings were elegant reimaginings of the thermostat (Nest Thermostat) and the smoke/carbon monoxide detector (Nest Protect). In addition to selling direct on its website, Nest makes its products available through retail partners, including Best Buy, Apple, Lowe's, and Home Depot. Matt Rogers, cofounder and VP of Engineering, was previously a firmware engineer at Apple, where he worked on the iPod. In this first of two case studies on the company, he discusses the importance of branding.

The Nest team began to create their brand identity long before they publicly announced the company or the product. "We knew we had to build a consumer brand from very, very early on," Matt remembers. "We made it a priority. I think a lot of companies in our space, and a lot of companies in general that are building for the first time, spend a lot of time on the product and the technology. The brand and how you're going to market it is just as important."

As evidence of the extent to which Nest prioritized branding, Matt points to the team's early hiring priorities. The first two hires were engineers. The third hire was Erik Charlton, currently the VP of Business, who was originally brought on to run sales and marketing. While Matt and the engineers worked on building and designing the product, Erik and Tony Fadell, Matt's cofounder, spent a lot of time on the story behind it. "You can build a great product, but if you can't explain what it is and how it works really simply, and then build a story around it, it's not going to sell," Matt says. The team thought about what the brand imagery would look like, how they would talk about the brand, and how they would portray their products.

Nest ultimately chose "unloved devices" as their way of describing the many products in the home that were traditionally overlooked. There are many thermostats on the market; the Nest thermostat would differentiate itself by being a *loved* product. To deliver on that promise, the team worked on creating a user experience and design replete with what Matt describes (*http://bit.ly/ ipod2thermostat*) as "that ineffable quality that makes people want to touch, hold, play with your creation." They prioritized clean design and smart functionality, and fought feature creep.

The brand and story are very important for selling a product to consumers. But having a distinct brand identity can give you a leg up with retailers as well, particularly for an early-stage startup. Big-box retailers are often wary of dealing with startups, because they prefer to avoid contracts for single-SKU companies with unproven products. Matt observes, "When you speak to retailers about a product and already have the whole story around it and how you're going to market it, it makes you much more impressive."

The Nest team made a strategic decision to approach Best Buy first. Traditional (unloved) thermostats are primarily sold at home-improvement and hardware stores; Nest wanted to be on the shelves of a consumer electronics powerhouse. The team's first conversation with the retailer happened soon after they started the company, before the product was even built. They wanted to gauge Best Buy's interest in selling a thermostat, something they had never sold before. "When we told them what we were working on they laughed," Matt says. "They said, 'Really? We don't even sell thermostats. Why should we be involved in this?'"

The team had anticipated this. Their first blog post—"Thermostats? Yes, thermostats." (*http://bit.ly/yes_thermostats*)—introduces the company by poking fun at the skepticism and then goes on to clearly articulate Nest's value proposition and differentiators: beautiful design, smart (it learns user temperature preferences and habits), remote-controlled via smartphone, and reduces energy bills. They pitched that vision to the representatives from Best Buy.

Although they kept their product roadmap secret from everyone (including retail representatives), Nest prioritized creating a brand mission—"We rethink unloved devices"—that was bigger than its initial product and would easily extend to other products. The founders applied values they'd developed during their time at Apple—clean design, attention to detail, delightful user experience—to every aspect of product development. The result is a product loved by many (the first run was sold out (*http://bit.ly/nest_sold_out*) for months) and a brand name synonymous with beautiful connected devices.

Now that you've found your team, identified your customers, targeted your market, and begun to formulate your brand, it's finally time to start building the prototype.

CHAPTER 5

Prototyping

Now that you know your user, market, and brand, you should probably get to prototyping. In fact, you've probably already been prototyping, whether you know it or not!

Many entrepreneurs think of a prototype as something that looks and works like the final product. The reality is that many, many prototypes of varying success and resolution are necessary during any hardware product development cycle. One phrase commonly used to describe this approach of iterative prototyping is "fail early and often." It is as true on the prototyping side as it is in business.

Reasons for Prototyping

The most important reason to prototype is fundamentally to *learn*. You should be learning from each prototype in one or many ways: showing it to potential users for feedback, demoing it to VCs to prove your concept, or even just learning about the prototyping process itself.

It's important to learn from every prototype by defining a *hypothesis*, or something you expect to learn from each prototype. This can help you decide how much time and money to spend on the prototype. If you're bringing a prototype to a VC partner meeting, don't cut costs: get it from a nice model shop. You also wouldn't want to spend too much time on your first prototype after your napkin sketch, simply to flesh out your idea in a physical form. It's important to consider these trade-offs in hardware, since prototype cycles can take 2 to 10 weeks from outside vendors, depending on the process you need and the complexity of your prototype.

In addition to learning, prototyping is also fundamentally a communication tool. This is often overlooked by newcomers to prototyping and can be hard to explain to someone who has never had to explain what a physical prototype means to them. One company accustomed to the burden of teaching clients the prototyping process while also delivering to them a final product idea is IDEO, a holistic product design and engineering company headquartered in Palo Alto, California. Dave Lyons discusses this further in "Prototyping, and Explaining Prototypes, at IDEO: A Case Study" on page 80.

Prototyping, and Explaining Prototypes, at IDEO: A Case Study

Dave Lyons is an entrepreneur with a deep background in engineering, design, and product development. He cofounded Peloton, an automated vehicle technology company, in 2011 and was previously director of engineering at Tesla Motors, where, as Employee No. 12, he led and expanded the development team for each of the company's core technologies through their formative stages. He also previously spent 11 years at design and innovation consulting firm IDEO in various roles. Dave subsequently worked as an entrepreneur-in-residence (EIR) at two Silicon Valley VCs. Here, Dave describes the challenges of helping clients through the prototyping process at IDEO, as well as some best practices for prototyping that he implemented at subsequent startups.

Dave first worked at IDEO from 1991 to 1999, a time when the importance of holistic product design was skyrocketing in Silicon Valley but still often needed to be explained to clients. Dave says that some clients walked in the door saying "just give me the answer, just give me the answer" in the form of a final idea or prototype. Dave says that those clients often needed a closer relationship to show IDEO's creative design process and make sure that all sides were excited about the final design achieved.

Something Dave took away from his time at IDEO was the importance of embodying your ideas in a physical prototype. Dave insists that every time you get people together, you should represent the concept in the form of one prototype or another. This is because "every time you bring a prototype or a model to a meeting, it is the center of the conversation." Dave says that prototypes are fundamentally made for communicating ideas, but prototypes alone were often not enough for communicating the vision to clients.

It is also important to note that prototyping was at a similar technological state then as it is now. Dave points out that, "In the middle 1990s, we had 3D printing, and the 3D printing was almost as good as it is right now; it just cost a ton more, so you didn't do it. We had 3D CAD (computer-aided design), and the 3D CAD was still 3D, it just didn't have the same rendering capabilities." This is important especially in light of the recent popularity of 3D printing and the proliferation of availability. Just because MakerBot has made fused deposition modeling (FDM) 3D printing technology accessible to any maker in the world, it doesn't indicate that the FDMs used by professional engineers have changed in capability or cost as much as it indicates an awareness by the public of this existing technology. This is important to consider, as new prototyping and manufacturing fads are bound to pop up in the future.

Dave says that another large benefit of prototyping early in the process is to "shake the gremlins out of the design as fast as possible in different areas," to figure out what isn't working, and then embrace a company culture of "willingness to look beyond the artifact on the table towards what it might be." While getting a prototype to the meeting is important to show the direction the design is going and to spur a conversation, it's also important to look forward to the potential new aspects that the design suggests and not to get caught up on small details of the prototype, such as the specific color or the edge treatment of a stock material used.

Dave also insists that it's important to prototype early and often and to craft your prototype to your audience: "If you bring it to a meeting of designers, you can bring rough foamcore. If you're bringing it to a meeting of venture capitalists, you want to make sure that you've brought something a little bit cleaner." Your audience for each prototype is very important to consider, as discussed in more detail in "Types of Prototyping" on page 83.

With Dave's recent work at Clover, a point-of-sale system, they routinely separated *looks-like* and *works-like* prototypes (see "Works-Like and Looks-Like Prototypes" on page 85):

> *We constantly brought models that were generally in the same form. We would bring renderings that were photorealistic, in fact, better looking than the real products, and we would also then bring works-like prototypes, mechanisms that might show the articulation of a display between a customer and user side.*

Having all three of these prototypes together—form studies, computer renderings, and functional works-like mechanisms—allows an entrepreneur to "find the right fit or leap of faith for the audience." This divide-and-conquer prototyping approach is also discussed later in this chapter.

Throughout multiple companies, Dave has always stressed the importance of prototyping as well as the importance of presenting your prototype to

> stakeholders, including users and investors, early and often to get feedback and to make sure you're going down the right path.

There is an additional benefit to low-resolution prototyping, beyond the saved time and cost. Often, showing a raw version of a prototype or idea will garner rawer feedback from your users. If you show users a fully fleshed-out prototype that looks like a real product, a user is likely to suggest only minor details about the color or button interface. Showing a raw prototype will get you more basic feedback about the product, including fundamental features.

Hand-waving and magic wands are fair game in early user trials. Often, you can make the first prototypes to show to users out of cardboard, foamcore, or even paper. Foamcore is a super handy material found at art supply stores for early prototypes consisting of a layer of foam bounded on both sides by paper. You are probably familiar with the material as the de facto display stands from elementary-school science fairs. You can mock up user interface (UI) screens from the mobile app, create rough shapes of the product, and make other low-resolution prototypes to explain your ideas to the user and refine the feature set of the MVP.

It can also be helpful to keep the fancy shapes and curves out of the design you're showing to users until you have an industrial designer take a pass at it. Stick with rectangular boxes and cylinders containing all the parts for the early prototypes. Having slightly varied prototypes will also help to do A/B feature testing quickly with users. It's easier to ask users about details that are fully embodied in a prototype; you can just ask the users to choose between prototype A or prototype B.

Prototyping can also help you learn about the capabilities and limitations of different processes. Especially if you're transitioning from a software or business background, it's important to get your hands dirty making. Go to TechShop (or a workshop or lab in your area) and learn how to use the laser cutter! You should also start to learn about manufacturing-scale processes with similar capabilities so that you can start to compare the trade-offs you will need to make further down the road. (See Table 5-2 in "Prototyping and Manufacturing Processes" on page 108 for specific trade-offs between mechanical prototyping and manufacturing processes.)

In general, it's important to continually increase your prototype resolution and keep moving your processes closer and closer to the final manufacturing process. This will help show progress to users and investors who have been exposed to your prototypes all along ("It's looking more and more real!"), and also help your team learn about the capabilities and limitations of manufacturing processes once you start to mold millions of parts instead of 3D printing one or two.

If your product requires an app or website, then so will you during the prototype stage. However, that doesn't mean that you need to have code to go with your foam model. Instead, you should start simple and focus on the interactions and user flows. There are many tools that allow for quick prototyping. Paper is one of the simplest. In just minutes, you can draw the necessary basic UI elements. The next step is to use wireframe apps like Balsamiq or POP that come with templates. Once you really know what you're doing and you want to test actual working code, you can use app-generators like the open source PhoneGap (*http://phone gap.com*) (code in HTML, CSS, and JavaScript) or MIT App Inventor (*http://appinventor.mit.edu/*).

Types of Prototyping

There are many approaches to developing prototypes for hardware products. Some engineers will jump straight into a CAD system to model the product on their computers. Others will spend weeks in the machine shop creating form studies out of foam before they touch a computer mouse. There is no correct approach to which tool to use first, and it is highly dependent on what type of product you are working on as well as your background.

For developing the types of consumer products commonly produced by hardware startups, we recommend starting with physical prototyping in a shop, which will allow you to quickly make your idea real in the world, get it out in front of users, and learn from their feedback. This is because the overall user experience (UX) and interaction with the product will likely end up being the key differentiator rather than a completely novel technology.

At some point, you will want to put your design into CAD, likely when you need to to start making 3D prototypes, but you probably won't need beefier analysis tools until you move toward manufacturing. Computer design software is usually divided into two main categories—overall

CAD, and electronic design automation (EDA or ECAD) tools specifically for electronics:

CAD
   Programs for designing mechanical or electrical schematics, parts, assemblies, and drawings. These programs are typically divided among tools for industrial design (ID), mechanical engineering (ME), or electrical engineering (EE). Common mechanical CAD systems for hardware product development include Dassault Systèmes Solid-Works and Autodesk Inventor or Fusion360.

EDA or ECAD
   Advanced board layout is done in software. Altium (*http://www.altium.com*) is a popular but expensive package. EAGLE (*http://www.cadsoftusa.com/*) by CadSoft is a popular freeware package that has plenty of community support and extensions.

Prototyping first isn't always the best approach. If your core differentiator is a new type of technology such as a sensor, it's important to prove that functionality, possibly through computer analysis, before embarking on a series of prototypes for a product that might not work. Also, going through the process of proving your technology might be a learning opportunity for your team that could propagate to other areas of the design. Maybe you need to introduce a new material or color to achieve functionality; that could subsequently become input for your industrial designer to include as a feature or accent for emphasis.

PROTOTYPING TERMS

When working in prototyping stages, you might come across some of these common phrases to describe elements of the process:

Proof-of-concept (POC)
   Used to prove out one tricky technical challenge and lower the risk of continuing development.

Gestural prototype
   A rough representation, usually used to explain a concept and garner feedback from users, meaning there is lots of hand-waving and saying things like "imagine if it did X" to visualize the device's user experience.

Breadboard prototype
   A phrase often used by mechanical engineers to indicate a build that integrates subsystems, although the phrase is rather confusing, because it can be conflated with an electrical engineer's *breadboard*, which is a flexible matrix of wire inputs used for prototyping.

Scale models
   Models smaller than the actual product can also be productive communication tools if the real product is too large to carry around with you to every meeting.

Whatever colloquial phrases your teams use to communicate prototype progress, it's important to define those early, so that everyone clearly understands the technical progress.

WORKS-LIKE AND LOOKS-LIKE PROTOTYPES

It can often be helpful to break down the scope or thesis of each prototype you embark upon, and even work on differently scoped prototypes in parallel. One common approach to this is the separation of *works-like* (WL) and *looks-like* (LL) prototypes. A WL prototype focuses on the core functionality of the product to make sure the technical challenges have been met and fundamental subsystems work before adding the burden of integration.

Examples of areas ripe for WL prototypes include sensor, mechanism, or connectivity solutions. An LL prototype focuses on the form, aesthetics, and design language, and potentially even touches on a product's ergonomics to emphasize its look and feel, regardless of its internal functionality. Dave Lyons refers to this approach to prototyping at Clover in "Prototyping, and Explaining Prototypes, at IDEO: A Case Study" on page 80, in which a company develops separate form studies, virtual renderings, and mechanism WL prototypes to explain the vision of the next product.

Another great example of LL prototyping is the collaboration between Apple and IDEO to develop the first computer mouse for consumers. IDEO made hundreds of form studies to establish the best ergonomics and look for Apple's first mouse through rapid prototyping, mainly in foam. While some aspects of ergonomics can be predicted through a CAD system, this iterative approach to prototyping shows that the true qualitative feel of a product and universal adoption by many distinct users can really be vetted out only by prototyping the forms and putting them into people's hands.

There are even more reasons to separate LL and WL prototypes during iterative prototyping cycles. From the design side, you can get user feedback on the form and feel of the product from users with an LL prototype, without the product necessarily working yet. Separate WL prototypes can help engineers solve smaller problems on a test bench before trying to integrate with a form factor or CMF (color, material, finish document) that adds additional constraints to the problem. It can also help disparate groups within a development team make progress in parallel, although the reality for startups is that all these responsibilities will often be embodied in one multitasking cofounder (probably the CTO).

It's also important to separate subsystems when prototyping complicated products. This can help during user trials, to isolate feedback to one area or feature set. It can also help in benchtop testing, to make sure one subsystem is working before integrating it with others. Mechanical engineering (ME) and electrical engineering (EE) prototyping is often separate at the beginning of a development cycle, with firmware and software integrating even later. That said, there is always a time and place for integration. Eventually, everything needs to work together as one beautiful, humming machine, to make you confident enough to release for tooling and truly begin your manufacturing process.

Sometimes, engineers can get stuck in *analysis paralysis*, unable to completely perfect a design, feature, or algorithm, and thus unwilling to release the current version and move on. It's important to know when to launch your MVP and move feature sets and improvements to version 2 later. You can also use feedback from the market on the first product, suggesting where to direct your engineering efforts for V2.

TEARDOWNS

Taking apart competitors' products already on the market (commonly referred to as *teardowns*) can be helpful early in the prototyping process. iFixit (*http://iFixit.com*) is the best resource for well-documented teardowns of common products. The site focuses on repairability and has clear pictures and documentation of the teardown process. Taking plenty of pictures and saving every part that comes out of the product can help capture what you learn from the teardown for the future.

There are a few important things to look out for while doing teardowns. One of the first is the component selection (especially high-value components such as the processor and battery) and part breakdown of the product, which can help you predict how much the device will cost to produce. It's also important to note how you think each part was manufactured, as well as how the product was assembled, so that you can learn from any unique approaches to manufacturing or assembly. Consumer products often cleverly integrate functionalities into a single part or use miniaturization methods, because there is such a burden on size for consumers. It's important to learn from teardowns, but don't simply copy your competitors' ideas.

Assembling Your Team

There are many different roles that need to be served during your development process. This section defines the most important disciplines for hardware development and gives some guidelines as to when you should engage which discipline, but it's important to emphasize whatever discipline lowers the current risks in your startup.

For example, it could be important to engage design services like branding, user experience, and industrial design if you need help crystallizing your vision and explaining the value proposition, but if your burden from investors or the market is to prove that your wacky idea will actually function, emphasizing mechanical engineering, electrical engineering, or firmware engineering early on may be more important.

INDUSTRIAL DESIGN

Industrial design (ID) is the process of creating the look and feel for a product or brand line. Industrial designers will usually take in a lightly defined scope, including the features, use cases, and target markets make sense for your product—including any known sizes or volumes that need to be accommodated for functionality—and will then determine what the product should look and feel like.

Once you've decided on your target market, branding, and the feature set of your MVP, you should bring your concept to an industrial designer to help give your product a look and feel. The normal output of these engagements will be a CMF document detailing the look and feel of the product, along with some form of CAD. The output can be a fully defined

3D product represented in CAD, a 2D hand sketch showing the vision, or a 3D model in CAD with slightly less information.

Normally, your industrial-design partner will not carry the design all the way to manufacturability, so a mechanical engineer, manufacturing engineer, or process engineer will need to get involved to make the tricky parts of the design real, or work with the designer to find an acceptable middle ground that can be produced at scale.

Prototyping ID concepts usually happens in the 2D methods discussed above first, then moves into 3D foam form studies. The most advanced ID prototypes are fully LL prototypes produced at a model shop with the capabilities to make them look like final manufactured products.

While industrial design can be the first step in large corporate *new product introductions* (NPIs) that have a clearly defined scope, it probably isn't the most appropriate first step for a cash-strapped startup. These services can often be expensive, so you will want to have a fleshed-out design brief with clear inputs and outputs before engaging firms for quotes. More and more often, industrial design is being used as a huge brand differentiator among many similar products, so it is important to consider early in your process.

USER EXPERIENCE, INTERFACE, AND INTERACTION DESIGN

Other design-related disciplines that you might consider engaging early in your process are *user experience* (UX), *user interface* (UI) design, and *interaction design* (IxD).

UX designers can help you zoom out to understand the entire experience a user goes through with your product, looking at the flow of use cases and design language among the physical object, mobile or web app, and any other software that adds to the product, to make sure you are delivering your user a cohesive experience.

Interaction designers can give similar help, but usually more on the software side of things than on physical hardware products.

UI designers can help you explore different interfaces for your product. Should a mobile app be the only way of interacting? Should the device have a local UI with at least red and green LEDs to assure the user that the device is functioning? Should your app have slider bars, graphs, or simple colors indicating trends? These are all topics that a UI designer can help you answer and then implement in your product. *Wireframes*, *screen renderings*, or *storyboards* are common prototyping tools for UI designers.

MECHANICAL AND ELECTRICAL ENGINEERING

ME and EE are often done close together with ID formgiving—the initial shaping of the product—or soon thereafter. Mechanical engineers work on mechanisms, enclosures, and the other physical elements that hold the product together, while electrical engineers work on the printed circuit boards (PCBs) and other sensors and actuators of the system. These disciplines should ideally still overlap with industrial design to make sure your ID partner comes up with a form factor that can deliver the desired function. Once an ID direction is established, ME and EE prototypes can begin to flesh out the technologies, sensors, and overall user experience that you hope to deliver with the product.

Mechanical prototyping usually starts with foamcore and cardboard, roughing out 2D shapes similar, or in parallel with ID. Stepping past the X-Acto knives, laser cutters make great low-resolution tools for cutting out 2D shapes and starting to work with more representative materials. It's important to prototype in 2D first, especially for working test prototypes and mechanisms, in order to move quickly through prototypes and get raw feedback from users. 3D printing is usually the next step in realizing your product vision, because you can quickly generate a physical 3D model of your CAD; more-robust prototypes can be made later with CNC machining or more manufacturing-oriented processes.

Almost all of the tools that we mention here can be found at your local makerspace and can be used for a monthly fee. To use them you have to become certified by taking one of their classes. It's definitely cheaper than buying your own!

On the electrical side of prototyping, it's also important to iteratively prototype with more-flexible platforms before moving into custom electronics and trying to miniaturize to the final form factor. Arduino, Raspberry Pi, and BeagleBone are popular platforms (see "Integrated Circuits" on page 94 for more details). Many prototypes will start on Arduino in order to hook up a sensor and play with a few I/Os. If you need more processing power, you should instead start with the beefier Raspberry Pi or BeagleBone. Note that unless you choose the same chip or hardware architecture, you'll most likely have to write all new firmware when you start spinning your own boards.

Often, *firmware* (FW) development will come in after ME and EE development have made some progress, but it's important to consider while choosing your board architecture. Firmware engineers develop the code (generally C) that lives on a device permanently and directly controls all actuations and reads all sensors. Firmware engineers will work very closely with the electrical engineering team, often in the same group. Firmware prototyping will use many of the same tools as electrical prototyping, although it can be helpful to add logic analyzers and more instrumentation to electrical hardware in order to debug firmware code.

SOFTWARE

Finally, there's *software*. The majority of hardware startups will require a software team. That team will be responsible for the mobile app and the service. However, at the prototyping stage you don't need to scale just yet. Depending on the product, you may be able to get by just proving that you can collect data from your sensors or that you can make your hardware device function with your chosen connectivity solution. Though we don't recommend it, many teams start developing the software only as the hardware is going through design for manufacture (DFM).

Whichever order you choose to tackle these disciplines in, it's important to strive for lots of overlap among the disciplines so that no group falls behind and you don't end up with requirements that are unachievable by another discipline.

Outsourcing Versus Insourcing

Any hardware startup should consider what level of prototyping it should be doing, internally or externally. While any company that calls itself a hardware startup will have the standard soldering irons, cordless drills, and workbench full of disassembled parts, investing in larger equipment can pay off in the long run.

If you just landed a large seed round, you might want to spring for a 3D printer or some PCB equipment to speed up your prototyping cycles. Having larger prototyping equipment inhouse can not only make things faster; it can also familiarize your team with the prototyping process and help them better design for the tools. One downside is the overhead required to maintain this equipment, as it can often break and put a hiccup in your cycle.

OUTSOURCING

The main benefit to using outside vendors is saving time, but there are other factors, such as reliability and benefiting from their expertise in processes they are much more familiar with than you are. If you can afford to wait for the parts to come back, it can be worth it to send them to an outside shop.

When releasing a prototype to an outside vendor, there are many considerations to make. The first is which vendor is appropriate to your prototype process, timeline, and budget. If it's the first time you've used a certain process, you should send out quotes to at least three or four vendors and determine the trade-offs among them. This inquiry is often called *RFQ* (request for quote).

Traditionally, MEs have created 2D drawings to communicate their design intent and acceptable tolerances, but this has become less popular with the advent of ubiquitous 3D CAD tools and more standardized prototyping tolerances. It is common now to send a 3D CAD transfer file (*.stp* or *.igs*) accompanied by a PDF that shows the CMF along with any tricky areas to prototype. EE releases typically involve *Gerber files, drill files,* and potentially any paneling input on how to nest the boards, depending on the vendor.

It's important to not only send a vendor the files that you want to be made, but also to include expectations about time and specific requests. You will typically receive a range of quotes, and you might need to make a trade-off between cost and time (e.g., Vendor A can fabricate your boards in two weeks for $500 and Vendor B can fab them in one week for $750). Sometimes you will be offered these trade-offs even within one vendor's quote, including an expedite fee to turn around parts faster. Once you establish a relationship with a vendor and give it more regular business, it should also be able to incorporate you into its internal planning and hopefully give you more visibility into its actual timing so that both sides can plan more appropriately for builds and releases.

INSOURCING

If you do choose to do more of your work in-house, you'll need to stock up on your own parts and components, along with the tools to assemble them.

Parts and components

You can find many off-the-shelf (OTS) solutions for parts and components. One example is the glut of Raspberry Pi cases on Kickstarter. There's no need to 3D print every screw you need when you can likely get the right part in the right material shipped overnight. McMaster-Carr (*http://www.mcmaster.com/*) is the go-to source for mechanical parts and components such as screws and other fasteners, material stock, and other valuable prototyping resources. It can often be helpful to order a variety pack of different screw sizes, O-ring sizes, and anything else you might need different sizes of to enable a smoother build. There are also local sources such as your hardware store, Grainger stockrooms, and more industry-specific supply houses. It can often be helpful to go to stores in person to source materials you need, to familiarize yourself with a product and learn from knowledgeable sales staff. One example of this in the Bay Area is Douglas and Sturgess, a supplier of casting materials whose knowledgeable sales staff can likely help guide your prototyping process.

On the electrical side, digiKey (*http://www.digikey.com/*) and Mouser (*http://www.mouser.com/*) are the go-to places to get electrical components delivered in a hurry, including discrete components, integrated circuits (ICs), and more complex sensors and lab equipment. Sourcing these components locally can be rather difficult, though. Your local hardware store can be a good source for larger electrical components (such as thicker wire, light switches, etc.), but it often lacks fine-gauge wire, solder, and other components geared more toward electronics prototyping.

Engaging with vendors that supply specific components or connectors early in your process can also help your prototype process. Not only can you get data sheets and other design information, but you can also order samples of their commonly supplied items. This can help to ensure that you are designing around the right parameters and that their components will work in your system. Many vendors will supply free samples if you show them you are serious about eventually shipping a product.

It's also important to choose the appropriate process for your prototype. The most important considerations here are usually time, budget, and the scope of what you'd like to achieve with each prototype. (See Table 5-2 in "Prototyping and Manufacturing Processes" on page 108 for a rough guide to the most common processes for different stages and materials.)

Ben Corrado, cofounder of engineering development consultancy Rigado LLC (*https://rigado.com/*) in Salem, Oregon, has a wealth of experience helping startups develop products and has seen these trends changing over the years. He cautions startups against a racing toward miniaturization to the final form factor early in development. "That would be my first advice to people," Ben says. "Make it bigger on the first prototypes. In many cases, form factor isn't going to be the most important thing when you're debugging."

Ben adds that prototyping modern electronics in-house has become difficult, because "parts have gotten so small that you really can't prototype in-house as effectively anymore. When you're putting down a .4mm pitch BGA, it's not practical to do it without machine placement and X-ray inspection to make sure there are no shorts."

Tools

These are some helpful tools to have on hand:

Portable drills
A necessity for any making. You might need to upgrade to a stand-alone drill press if you are drilling larger objects or need repeatable hole placement.

Bandsaw
A saw with a continuous metal blade for cutting through wood, plastic, or metal, depending on the blade type. This is the most common saw for some detail cuts.

Table saw
A working table with a circular saw coming out of the center, usually used for making larger cuts of thin sheets, especially wood. Can be very dangerous.

Mill
A machine with a vertical cutting piece; handy for reworking parts.

Lathe
A machine that spins the work piece into a cutter in order to machine cylindrical profiles.

Soldering station
A handheld heating element used for heating up solder to melt components to boards or rework components or solder joints that are not correct. A solder iron has differently shaped metal tips for heating

components through touch, whereas more advanced solder stations have forced hot air for heating entire areas of a board.

Multimeter
The most basic electronic measuring device. Can measure many properties, such as voltage, current, resistance, and other values on more advanced devices. Normally portable, to fit in one hand, with two test leads that extend out with wires. Also known as a *digital multimeter* (DMM).

Oscilloscope
A more complicated electronic measuring device capable of showing not only the raw values of electrical properties, but waveforms as well. More complicated 'scopes move into the territory of spectrum analyzers and network analyzers that can analyze antenna signals and help lower risks during certification.

Pick-n-place
The common term for surface-mount technology (SMT) placement equipment, as the machine literally picks up individual components, normally through vacuum suction, and then places them in their correct location on a PCB.

Reflow oven
A small oven that heats up the entire board with components on it and solders them all into one printed circuit board assembly (PCBA). Some scrappy startups use an old toaster oven for a reflow oven.

Integrated Circuits

Electronics are the brains of connected hardware. They collect the data, make it useful, and share it with the users. When you first start prototyping, you often just need to make sure that you can gather meaningful data.

The starting point for many is the Arduino (*http://www.arduino.cc/*), an open source microcontroller whose software environment (the open source Wiring (*http://wiring.org.co/*)) can be programmed with Linux, Mac, and Windows using many popular programming languages such as C/C++, Processing, Python, and JavaScript. There are thousands of projects and code samples online to learn (or borrow) from. Accessibility has been a mainstay of the Arduino since its original release in Italy for students.

The explosion of Arduino projects has brought about a number of other prototyping boards. The Raspberry Pi is a Linux computer on a board; needless to say, it's much beefier than the Arduino! It has 256 or 512MB of RAM and USB ports. Another popular option is the BeagleBone Black. It also runs Linux and has ports, RAM, and storage. These are great boards that are inexpensive for prototypes and personal projects, but they generally cost too much when you start thinking about shipping product.

After one or two prototype versions, you will need to move beyond these dev boards and start spinning your own boards. At this time, you'll need to make an important decision whether or not to use a *microcontroller* or a *microprocessor*. Generally, the final firmware will be in C, or sometimes C++ (though this will take up more room on the device). To make the decision, you'll have to evaluate trade-offs among processing, power consumption, cost, and size.

A microcontroller is smaller, so it is good for wearables. It costs less, which is great for price-sensitive products. It also consumes less power, which is excellent for devices that rely on batteries. However, it also has less onboard processing power, which means it's good for data collection and passing data through, but not for running algorithms on the device.

You'll need to pick a real-time operating system (RTOS) to use. There are many proprietary ones, such as Nucleus or Integrity, and open source alternatives, such as FreeRTOS and µClinux (a slimmed down Linux variant). The Spark Photon is a great board for prototyping, if you know that you'll be using a microcontroller.

The Pebble watch (see "Pebble Watch: A Case Study" on page 176) is a great example of the flexibility of a microcontroller. The original Pebble has a 144 × 168-pixel black-and-white e-paper display; Bluetooth 2.1+ EDR and 4.0 (Low Energy); four buttons; a vibrating motor; and a 3-axis accelerometer. The Pebble also supports third-party apps and watch faces, all of which run on a microcontroller with 128k of RAM and 512k of program space (by comparison, the first watch, InPulse, had 8k of RAM and 32k of program space) and a battery that lasts a week. As Eric Migicovsky, CEO of Pebble, notes, "When you think about it, [the Pebble] is just as powerful as the original, classic Mac, but on your wrist."

The Pebble's microcontroller runs FreeRTOS (*http://www.freertos.org*), an open source RTOS designed for devices. It provides the backbone for the software on the Pebble, but the team built everything else. According to Eric Migicovsky, "We built everything from the drivers to the firmware to the protocol between the phone and the watch. We built the Android and iOS apps, we built a cloud compile system. Everything, pretty much."

A microprocessor has a lot more processing power than a microcontroller. It has its own onboard RAM and flash, so the cost and power consumption will be higher. It can run a full operating system such as Linux or Android, and even higher-level languages like JavaScript. Many companies use Yocto (*https://www.yoctoproject.org/*) to create custom real-time Linux distributions for their products.

In many ways, the gaps between microcontrollers and microprocessors are closing. MicroPython (*https://micropython.org/*) is an almost full version of Python 3.4 designed for microcontrollers. The creator used Kickstarter to produce WiFi-enabled boards. Tessel (*https://tessel.io/*) is a new microcontroller that runs JavaScript. It's not the same as having full Linux onboard, but if you don't have a firmware resource on the team, then at least you can get the board out the door.

After you've decided what to use for the brains of your hardware, you need to get the physical chips. The best place for the necessary information is often the chip manufacturer's website or your local sales rep. You can get the reference designs and learn about their tolerances, and acquire dev kits. Companies like TI, Freescale, and ARM will send you dev kits for free (or at least a low cost). They want you to work with their chips at this stage, because they know that this could turn into orders in the hundreds of thousands (if you're so lucky!). The thing to realize when working with dev kits is that they don't always reflect the "real world." They are designed to let you work with anything, so they might have more RAM or higher tolerances so that you can use any power system.

Next, you'll need other components, such as antennas and sensors. You can also get samples of these from the manufacturers. However, you have some decisions to make that will balance cost, power consumption, performance, and physical size. Will you build your own? Or use an existing module?

Modules are physically larger and cost more, but they can save you time; because they are guaranteed to work and are already certified, you can skip checking your intended radiated emissions. However, you will still need to test for your unintended emissions certification (FCC class A or B—more on this in "Certification" on page 134). Modules can be useful if you are developing a custom radio frequency (RF) or some other highly specialized function and don't have a lot of experience in that area or a lot of time.

You'll have to be careful when picking a module. You'll be constrained by the vendor's choices; that is, you will have no ability to customize, and they might be too big or have a low-gain antenna. Ultimately, developing your own IC is cheaper and they are physically smaller, but the design time and nonrecurring expenses (expenses incurred only during the initial development of a project) are higher. One twist on this model is the Spark Core (and new Photon). Spark makes its boards and firmware simple to transition from prototyping to manufacturing. More and more teams are using them. Learn more about their story in "Spark: A Case Study" on page 97.

Spark: A Case Study

Zach Supalla is the CEO of Spark (*http://Spark.io*), a platform company that is fast becoming the development kit and cloud service provider of choice for hardware startups creating connected devices. However, like so many companies, it started out doing something (sort of) completely different. Zach's father is deaf, and Zach wanted a way to make the lights in his father's house flash when his mother sent him a text message. So they launched a Kickstarter for the Spark Socket (*http://bit.ly/spark_socket*), but the campaign was unsuccessful. Zach remembers, "We were asking for $250,000. We only raised $125K—enough to say that there was something there, but not enough to actually manufacture. And of course, Kickstarter is all-or-nothing, so we got zero dollars."

Failure was Spark's lucky break. Its initial exploration led it to realize just how hard it was to create connected products. "We learned firsthand that there were a lot of unforeseen complexities, just things that were hard that we weren't expecting to be hard," Zach says. "And we thought, 'Well, there is probably a lot of people out there who want to make connected products and run into these same problems.'"

So Spark decided to build infrastructure that would enable others to connect hardware to the Internet. It took the brains of Spark Socket and rejiggered it into a development kit called the Spark Core. Thus its first shipping product was born: the Spark Core, an Arduino-compatible, WiFi-enabled, 72 MHz ARM Cortex M3 board. It brought it back on Kickstarter, and this time it blew past its modest goal of $10K and raised $567,968.

It has since shipped 50,000 Spark Cores and raised a Series A venture capital round. Spark has also announced an upgrade, the Photon, with increased memory and a Broadcom WiFi chip, as well as a second product, the Electron, which provides development tools and infrastructure for cellular-connected products.

Nomiku (*http://www.nomiku.com/*), a fellow alum of HAXLR8R (*http://www.haxlr8r.com/*), is one of the first companies to both prototype and ship with the Spark. The Spark platform came in handy when Nomiku added connectivity in the second version of its *sous vide*.

According to Zach, "A ton of our customers are hackers and hobbyists who are just screwing around and making cool stuff, and posting stuff on the Internet. One of them was Warkitteh, which was a cat collar that would basically scour the neighborhood for open WiFi networks because it's on the cat's collar."

Now Spark is expanding into web infrastructure to support more-complex products. It has decided to create a full vertical stack that links its hardware to web services: Spark OS. Spark OS will offload the processing that can't be done on microcontrollers and handle securely sending back the relevant data for the user. This allows for a lot of use cases that might seem simple but aren't. Let's assume that you want a lighting system that turns on when the sun goes down. Zach says:

> *For the device itself to do that is actually kind of tricky, because it has to know what time it is, what day it is, what time of the year, and what time zone it's in, so that it can figure out when the sun is going down. Whereas you can do that in the cloud and have one application that just knows all that stuff and then sends out messages to all of the lights that say, "Okay, you turn on, you turn on, you turn on, you turn on" as the sun goes down wherever each device is located.*

By intimately tying the processing back end to the hardware, Spark is attempting to pave the way for thousands of nontech companies to bring their products online.

Connectivity

There are many reasons for connecting your product to the Internet. You'll want to be able to upload and download data, send commands, and update your device. If you've used a low-powered microcontroller, then any advanced processing of your data will be done on your phone or in the cloud. If you don't have experience with antenna design, it's going to be much faster for you to use a precertified module.

The connection can be done directly (WiFi, cellular), through a hub (usually ZigBee or Z-Wave), a phone (Bluetooth, WiFi), or a cable (often via the same port used for charging). All of these are proven models, so it really comes down to what scenarios you want to be able to support.

Hubs enable flexibility and are useful when you are working with a family of products. For example, SmartThings (an IoT company acquired by Samsung) requires a hub. The hub connects to the Internet via an Ethernet cable, and all SmartThings products use ZigBee to connect to the hub. The SmartThings mobile apps just connect directly to their back end to control the system and read the data.

When making your connectivity choice, you must consider the following issues:

Location
   Can it be hardwired to an Ethernet port, or do you need a wireless option? Will it be within a few feet of a person when in use, or does it need a broader range?

Movement
   Will the product be stationary or on the move? If it's on the move or can't necessarily be located near an Ethernet port, then you'll need a wireless option.

Power consumption
   Can the product be plugged in? Will the user expect constant connectivity? Will the user mind charging it frequently?

Cost
   Is the product expensive? Some connectivity options are more expensive than others. Cellular service often requires a subscription fee.

Data
   How much data will your product be transporting, and how quickly does it need to happen? WiFi and LTE have high data-transfer rates. All of the other options are an order of magnitude slower. Some (ZigBee and Z-Wave) are so low that they are best used only for control data.

Table 5-1 illustrates the trade-offs among wireless connectivity options.

TABLE 5-1. Wireless options [a]

Wireless technology	Frequency	Typical range	Power consumption	Peak data rate
WiFi (802.11n)	2.4GHz	300ft	Med	600mbps
Bluetooth 2.1	2.4GHz	30ft	Med	3mbps
Bluetooth Low Energy (BLE); Bluetooth Smart	2.4GHz	30ft	Low	1mbps
Cellular (LTE)	700, 850, 1700, 1900MHz	>5miles	High	1gbps
ZigBee	2.4GHz	60ft	Low	250kbps
Z-Wave	900MHz	50ft	Low	100kbps

[a] Table 5-1 courtesy of Brian Lee.

Here is a deeper description of the technologies described in Table 5-1:

WiFi

WiFi has a long range and high data throughput but consumes more power. Often the device connects to a WiFi network. In general, WiFi devices should be plugged into the wall due to their high power consumption. There are some battery-powered products, like cameras, that generate their own WiFi network so that they can sync lots of data quickly to a mobile device.

Bluetooth

Bluetooth has low energy consumption and thus is incredibly popular for wearables and other battery-powered devices (especially the newer Bluetooth Smart, a.k.a. Bluetooth Low Energy or BLE). However, this is coupled with low data-transfer rates. Also, Bluetooth is notoriously finicky to get working on mobile apps.

There is an ever-increasing number of tools for BLE. The San Francisco–based Punch Through Design, LLC (*https://punchthrough.com/*) has released a set of them. The LightBlue Bean (*https://punchthrough.com/bean/*) is a module powered by coin-cell batteries that can be used for prototyping. It has the advantage of being programmed wirelessly via LightBlue (*http://bit.ly/lightblue_ble*) and its iOS app.

Cellular modem

This connects to the cellular infrastructure to deliver location or transmit data and can enable data transfer and positioning even while outside WiFi range through GPS (Global Positioning System). GSM (Global System for Mobile Communications) and CDMA (code division multiple access) are the most common technologies. This is the most expensive option due to patent load on the chips, and (usually) requires a monthly subscription fee. If you have a high-priced product or a subscription, then this might not matter, but if you're aiming for a relatively low cost, this can add a significant cost to your bottom line. A great example of this is the Kindle, Amazon's ereader. The WiFi-only version is generally two-thirds the price of the WiFi +3G option. Amazon is baking in the additional cost of the 3G chipset and subscription fee. If the product is not going to be within consistent WiFi or BLE range, it might be the only choice. Spark (discussed in "Spark: A Case Study" on page 97) launched a successful Kickstarter campaign for the Spark Electron (*http://bit.ly/electron_kickstrtr*), a cellular dev kit with minimal data fees, in March 2015.

ZigBee/Z-Wave

ZigBee (*http://www.zigbee.org/*) is an open protocol; Z-Wave (*http://www.z-wave.com/*) is proprietary. They are often lumped together because both are good for short-range, low-power-consumption, low-bandwidth use cases. They are found in many IoT products (such as Smart Things' hub), but that may change as Bluetooth's popularity rises.

And here are a couple other radios of note:

Near Field Communication (NFC)

NFC is a short-range (10cm) radio that is available on many mobile devices. It can be used to sense the presence of and acquire data from NFC tags or peers (other phones). Many Android phones have an NFC radio (check this List of NFC-enabled mobile devices (*http://bit.ly/nfc_enabled*)). Apple has recently included an NFC reader in the iPhone 6, iPhone 6 Plus, and the Apple Watch, but has restricted its use to Apple Pay only.

iBeacon
iBeacon is Apple's proprietary Bluetooth-based technology. It functions similarly to NFC in that it is inherently a location-based technology. You can learn more about its intended use cases on Apple's iBeacon for Developers page (*https://developer.apple.com/ibeacon*).

Whatever connectivity and radio options you choose, make sure they fit your product's needs. Ideally, this part of the product fades into the background once you get past initial setup and becomes just a portal for your software.

Software Platforms

Even though this book is focused on hardware, you can't escape software. Almost any startup that is able to get funded will connect to another device or directly to the Web. Consumers expect the convenience of control. VCs want the promise of software-like scale. You'll want the ability to add new features, kill bugs via updates, and know more about your customers and potential revenue from apps and data.

By the time you launch, you're going to need a team with expertise in the following:

Interfaces (and gateways)
Most devices come with a mobile app. The app is used to configure the device, sometimes to connect to the Internet, and sometimes to serve as the screen for viewing historical (or even current) data from the device.

Backend
Many connected consumer devices get their computing power from large server farms. The device collects the data, and the backend service processes it and connects the data to your account.

Firmware
This is the code that actually sits on your hardware. It controls the power consumption and the connectivity. It is made up of integration code (generally from vendors), OS libraries, and increasingly, embedded Linux or another real-time OS such as FreeRTOS. It's software,

but because of its direct interaction with the hardware, it is often handled by the hardware team rather than the software team.

Website

A company's website is incredibly important. For hardware devices, these are often marketing, commerce, and support sites. Selling from your own site will get your highest retail margins and allow you to describe your product in your own words. Though an associated app might exist on the same domain, we are not referring to the design of the app here.

Software is different from hardware. While every software user gets the same bits (assuming you've done a good job with your source control!), every hardware user gets a unique set of atoms. Software builds on top of existing software, and what you prototype can often be turned into a shipping product.

When choosing your platform, our advice is to go with what your team knows, determine what is adaptable to your desired feature set, and make sure you can scale for the first set of users.

You'll want to factor in what your teams know. For your backend, Java might be where you want to end up, but if your team already knows Rails, a Rails app will be much faster to develop and ship. Will Rails have to last forever? No, but it should it be good enough to get you through your first tens of thousands of users.

There are many handy libraries that can be used for both prototyping and production. The Hybrid Group (*http://hybridgroup.com/*) has open sourced Artoo (*http://artoo.io/*) (for Rails development), Cylon.js (*http://cylonjs.com*) (JavaScript for in-browser control), and Gobot (*http://gobot.io/*) (for Go development). These libraries enable quick access to a variety of components (such as cameras and buttons), drivers (*http://artoo.io/documentation/drivers*), and devices (such as Leap Motion (*http://artoo.io/documentation/platforms/leapmotion*) and Arduino (*http://artoo.io/documentation/platforms/arduino*)). For really quick prototyping, it's often worth using an existing product. These libraries can control quite a few popular products (including Pebble (*http://artoo.io/documentation/platforms/pebble*) and the Roomba (*http://artoo.io/documentation/platforms/roomba*)).

On the mobile front, the native toolchain is much clearer. iOS devices require development in either Swift or Objective-C. Android apps are typically in Java. For faster development, you could use a multiplatform environment such as PhoneGap to produce the same app for both

Android and iOS. There's a clear speed advantage in development when you're producing two apps at once, but you should still factor in testing for each platform. This is a clear win for prototyping—as long as those are the platforms you want to ship.

There are additional considerations beyond development time. By using a multiplatform framework, you're sacrificing performance and UX conventions of platform (Android, for example, has a menu button, while iOS does not, and the platforms continue to diverge from there). Given the increasing emphasis on design, this might not be the wisest choice for shipping product. You could choose one platform to always be ahead and take turns developing for each. However, a startup with limited resources can make some compromises, such as developing one native app and then using PhoneGap to wrap a mobile website.

You'll want to determine what the load will be on your service. When you're dealing with hardware, most of the data is very small. Unless you are streaming data to the devices, most of them won't be on the system at the same time, so your calculation of concurrent users will be much lower than it would be with a software product. If you're coming from a web background, your experience will prepare you well, but you'll need to figure out how to use your existing tools with a hardware twist. In "Ringly's Software Development: A Case Study" on page 104, Tim Mason talks about his experience building Ringly's mobile and web apps.

Ringly's Software Development: A Case Study

Tim Mason is the San Francisco-based CTO and early advisor of Ringly (*https://ringly.com/*). He departed Etsy and joined the startup in January of 2014. Ringly sits at the intersection of fashion and technology. Its rings connect to your phone and notify you when something occurs that you might care about (such as an important phone call or text). The notifications come in the form of buzzes from a vibration motor and a series of various colored LED flashes on the ring. The exact number and color are determined by the user, based on the app or person that's trying to get his attention.

Like many great teams, Tim and New York–based Christina Mercando, CEO, had worked together previously. Like many startups, their idea began when Christina encountered an issue in her personal life and wanted to solve it. In the spring of 2013, Tim recalls:

> Christina and I would brainstorm startup ideas whenever we saw each other. Then one time Christina told me she had this idea of having some alerter, because she's missed so many calls and notifications by having her phone in her purse. So she was taking some hardware classes to build prototypes. I loved it [Ringly] immediately. We decided not to pursue any other ideas. So next time, she showed me the designs for an app, and even though I was still at Etsy, I said, "How about I take a couple of weekends and throw this app together?"

On the basis of the idea, team, and prototypes, Ringly received funding from First Round Capital (*http://firstround.com*) and A16Z (*http://a16z.com*) in September 2013. The team moved from New York to San Francisco in October 2013 for the inaugural class of Highway1 and set to work building a looks-like, works-like prototype (see "Works-Like and Looks-Like Prototypes" on page 85) of their first product.

Tim had lots of experience developing for web and mobile. He had cut his teeth on web development at Organic, one of the first big web consulting firms, in the 90s. At Lucasfilm, he had worked in Java. At Wesabe, an early financial startup built on Rails, he developed his first iOS app. At Hunch, he continued working on mobile. While at Etsy, he ran the mobile development team and built its first Android app. However, he had never worked with hardware, so he focused on Ringly's mobile app and, with funding in the bank, he hired a contract firmware developer:

> Our initial steps were building a board and then connecting the prototype app. I added Bluetooth functionality through the app and then I connected to it from the board. We looked at functionality of the app and started notification studies. We made sure that the ring would understand the commands that were sent over. So we have a series of commands for settings. We tested brightness, vibrating motor, and how long to spin the motor in intervals.

However, getting Bluetooth working meant selecting the BOM (bill of materials) components in tandem:

> We were still doing some hardware research, trying to figure out which chip we were really going to go with, what vibrating motor we were going to go with, what battery, etc. The prototype was just

printed plastic with flexboards. We started with one microcontroller that came with a lot of libraries that got us up and running quickly. It would work, but we were having trouble with Bluetooth, so we started looking for a new chip. At this time, Nordic (https:// www.nordicsemi.com) had just come out with the nRF51822, so we got a few of those and tested them out. Its Bluetooth stack seemed really solid, and you could connect quickly. The contractor that we had, his term was pretty much up, and we had found another contractor who had significantly more experience doing Bluetooth.

We would talk about the functionality that we needed. We would scope out the command set that I would send over from the mobile app. He'd say, "Okay, I have this build ready. I'm going to send you a hex file." I would upload the hex file using Nordic tools on the phone. First, I would flash it through the computer, but then Nordic has this really cool iPhone app, where I could just take the hacks and upload the latest application with the feature set that we had agreed on, and then I would code against it. At the same time, I would set up our latest app.

So there was this exchange of us setting milestones for, "Okay, now we're going to add the notifications" stuff, and "I need this from ANCS [Apple Notification Center Service] in order to be able to do stuff on my end." He would send that over, I would write code to parse that, I would send over my commands, and he would send over code to parse that.

Lots of back and forth. We'd both have our own protocols that we'd send back and forth that way. I think we have more back and forth than normal setups for wearables, where a lot of it is stuff gets stored on here and then you just download the data. We're doing very small protocol calls. Like the smallest possible things that we can send over to save the battery.

Another big task that Tim faced was how to structure the notification data for a space- and processor-constrained device. At that time, the Nordic nRF51822 had only 16k of RAM, and that was going to add a lot of constraints. All notifications on iOS are handled by the ANCS. For security reasons, iOS apps aren't able to see other apps' notifications, but devices can. So Tim developed a system whereby the ANCS streams all the notifications to the ring, which in turn streams them back to the app, which then decides which notifications are important enough to activate the ring's vibration motor and/or LED.

"So I started working with ANCS, the iOS notification service, and that is how you can get notifications from the phone to the ring," Tim notes. "Essentially, the phone is the central manager and using Bluetooth Low Energy, devices can connect to the phone through the service on the phone called ANCS." It will then broadcast notifications to that device.

It would be a lot simpler to decode the notification stream on the ring. However, with only 16k of RAM on the device, there's simply no room to process the data. Tim and his team have had to come up with lots of smart ways to decide which notifications get priority and what to do if you are away from the ring for a while. "What happens is, if you walk away from the phone, and you disconnect and let's say you get 20 or 30 notifications, we try and be smart about what app it came from," Tim notes. "So if you get 20 or 30 from the same app, we manage the notification queue differently."

To figure out how to improve the rings, Tim has implemented Mixpanel. "We use analytics to keep track of failures. It helps us determine what bugs we need to address and which aspects of the device or software might be causing them. We segment by firmware build and production batch (DVT, PVT, etc.) among other attributes." It helps the team find issues that occur only in the real world.

Ringly has since shipped units and has released both an iPhone and Android app. However, the software development is never done and they are in the process of releasing a firmware update to improve battery life and allow for simpler over-the-air firmware updates.

Software Security and Privacy

News about security and privacy breaches have become commonplace, but it's a bad day when you realize that it's your product on the front page of TechCrunch for exposing your users' data. Take, for example, the tens of thousands of vulnerable Internet-connected cameras whose owners have no idea their lives are exposed because the cameras all shipped with the same default password. By contrast, Dropcam, the connected camera service acquired by Google's Nest, has security baked into the product. To access your Dropcam's feed, you need an account, and the video stream is encrypted.

When considering security, the first step is to think about the threat model. Which aspects of the system do you care about most? In the case of Dropcam, it knew that its users would care a lot about the security of their video feeds. In the case of one of Eric Stutzenberg's firmware clients, "They in particular did not necessarily care about the data itself. But they did care about unauthorized access to the device, to get the data. So for instance, there you're talking about just user authentication and ensuring that a user is authenticated prior to allowing them to get the information from the machine."

Here are a few additional considerations:

Firmware encryption
  Most chips come with encryption options. It's worth using these to protect your proprietary algorithms as well as to protect your devices from being rooted.

Traffic encryption
  Encrypt the data when transferring. Bluetooth has encryption options baked in. If the device is connecting directly to the Internet via WiFi or LTE, you can use HTTPS for a secure connection.

User authentication
  Your customers are used to creating and using usernames and passwords for services. Make them a part of the system. Always make sure that the user is authenticated before transferring data. It may seem obvious, but make sure that users can always change their passwords.

Once you ship your hardware, you will have to support it as long as your customers keep using it—often, much longer than you expected! Luckily, your software is much more malleable and can be changed frequently—even daily.

Glossary of Terms

This section contains definitions for common processes and components that you will come across during your hardware journey.

PROTOTYPING AND MANUFACTURING PROCESSES

During prototyping and manufacturing, you will likely encounter the following processes:

Milling
> Performed on a mill, this process involves fastening down the workpiece and using rotary cutters, most of which are referred to as *endmills*, to remove material.

Turning
> This process, performed on a lathe, involves spinning the workpiece and using a nonrotary cutter to remove material. Normally used for cylindrical parts.

CNC (computer numerical control)
> Automated prototyping or manufacturing machinery; normally refers to a mill or lathe. Usually classified by number of degrees of freedom controlled by actionable axes (e.g., 3-axis or 5-axis).

3D printing
> A range of techniques for building up a part layer by layer, sometimes referred to broadly as *additive manufacturing*. Most commonly uses the STL (stereolithography) file format.

Molding
> Normally, injection molding is used for plastics in consumer products. *Overmolding* (multiple plastics), *compression molding*, and *transfer molding* are other molding examples.

Casting
> Similar to molding, but normally done with metals not under pressure. Die casting, investment casting, and pressure casting are a few types.

Stamping
> Cutting and bending a piece of sheet metal into a shape. Single-stage dies, progressive dies, and progressive rolling are a few types.

Extrusion
> Pushing a material through a die to create objects of fixed cross-sectional profile. Many stock materials are produced through extrusion.

Die cutting
> Normally used on packaging, paper, and other thin sheets. Cuts out a 2D pattern.

Laser cutting
   Uses a concentrated laser to cut through plastics, wood, or thin sheets of metal.

Water jetting
   Uses a blast of water and grit to cut thicker metals and plastics.

Table 5-2 provides a reference for the most common fabrication processes for parts, based on material and dimensions.

TABLE 5-2. Appropriate processes for scale

	Plastic	Metal	Wood
2D Prototyping	Laser cutting	Water jetting	Laser cutting
2D Manufacturing	Molding; Extrusion	Stamping	Laser cutting; CNC routing
3D Prototyping	3D Printing (FDM, SLA); CNC Machining	3D Printing (SLS, DLMS); CNC Machining	CNC machining
3D Manufacturing	Molding	Casting	CNC Machining

The descriptions in Table 5-2 assume prototyping for a run of 1 to 10 units and manufacturing for a run of more than 1,000 units. Midsize runs are less straightforward; in-between ranges will typically end being very expensive per piece and will require some compromises between one strategy or another. The reference here is meant merely a general starting point, as budget, look/feel, and time can drive process choices.

ELECTRICAL COMPONENTS

These categories cover common terminology for electrical components:

ASIC
   Application-specific integrated circuit. Chip optimized to one specific application.

BGA
   Ball grid array. An SMT component with a grid of solder balls that provides the most interconnections possible. Inspection must be done with X-rays, and rework is very challenging.

Breadboard
: A board used to prototype electronics, generally with larger parts than you'd use in production.

PCB
: Printed circuit board. The backbone of electronics assemblies. Layers of copper and nonconductive plastic designed to support components as well as make the correct electrical connections. Typical subtractive manufacturing methods involves creating the copper layers through silk-screen printing, photoengraving, or PCB milling.

PCBA/PCA
: Printed circuit board assembly. A printed circuit board that has been populated with components. Involves running PCBs through a pick-and-place machine to place components, and then a reflow oven to melt the solder. *Wave solder* and manual insertion and rework are also common processes.

QFN
: Quad flat no-leads. An SMT component with no leads extending beyond the package footprint. Can be smaller components, but harder to debug and rework since the leads are inaccessible.

SMT/SMD
: Surface-mount technology or surface-mount device. This is the most common modern electronics manufacturing process.

Through-hole (or thru-hole)
: Electrical components with axial or radial leads. Formerly the most common form factor for all electrical components. In modern usage, these components tend to be larger and need more mechanical mounting consideration (e.g., capacitors, transformers, and other power electronics components).

SENSORS

This section describes common sensors that you will need for your product. They all come in many form factors and can often be found combined in one component:

Accelerometer
> Detects acceleration of a device. This is used in mobile phones and other devices to detect the direction of gravity and relative orientation of an object. This is often used in wearables to detect motion.

Gyroscope
> Determines the orientation and rotation of a device relative to a fixed frame. Often paired with an accelerometer to make positioning more accurate.

Magnetometer
> Detects the direction of a magnetic field. Often used relative to the earth's magnetic field if not near artificial sources. Triangulated with an accelerometer and gyroscope, this can give incredibly accurate positioning.

GPS receiver
> Enables the device to locate itself on Earth via the Global Positioning System (GPS) satellite network. It generally requires an unobstructed view of the sky (and thus has issues in cities with tall buildings). On cell phones, it is often supplemented with a cell tower or WiFi for more accurate geolocation.

Switch
> The simplest of input devices, switches come in many varieties, such as *toggle, temporary, rotary, tilt,* or *limit. Capacitive* switches that don't require pressing down are also becoming more popular, but they require more power than a traditional switch.

Temperature sensor
> Used for controlling the system, turning on cooling devices, or detecting changes in the surrounding environment. Thermistors, thermocouples, and resistance temperature detectors (RTDs) are the main classes of temperature sensors.

While these definitions cover some of the basic sensors used in hardware products, there are sensors out there for any application. Other sensors include relative humidity or pressure sensors, microphones, cameras, and load cells. Emitters of signals include LEDs, screens (LCD, OLED, or LED matrices are common), and speakers.

CHAPTER 6

Manufacturing

AT SOME POINT, ALL SUCCESSFUL HARDWARE STARTUPS WILL HAVE TO FACE the challenge of manufacturing their products at scale. Manufacturing involves a bit of a different mindset than prototyping. Prototyping is full of experimentation and invention, while often burning money. Manufacturing is about process and, hopefully, eventually *making* money. Prototyping is making *one* of something; manufacturing is making something that can be made over and over again by people you don't know.

This can be a hard and expensive transition for a young company with limited funds. Often, schedules slip, which means that it takes longer to gain revenue from selling the product. Also, *working capital*, the money used to fund inventory, is necessary to get the product manufactured.

This chapter is a companion to Chapter 5 and should not be read in isolation. During your prototyping process, you will make many technical and business decisions that will ripple throughout manufacturing, so we have chosen to cover those in Chapter 5, to make sure the issues are addressed early enough.

Here are a few terms you will find used to describe different kinds of manufacturers:

Original equipment manufacturer (OEM)
A company that makes the original parts, components, or assemblies that go into a product.

Original design manufacturer (ODM)
A company that specializes in manufacturing products for other brands. This approach is often known as a *white label*, since any company could approach the same ODM and sell the product under its own brand.

Contract manufacturer (CM)
: A company that handles the manufacturing and potentially the supply chain for another company. A form of outsourcing.

Electronic manufacturing service (EMS)
: A company that designs, tests, and manufactures electronic components and assemblies.

It's important to note that these terms have the potential to be used somewhat interchangeably, depending on geography and industry.

Preparing to Manufacture

In your process, you'll need to determine the best time to engage factories and CMs. If you approach factories too early with concept sketches and vague ideas, you might scare them away from your business, since they might not take you seriously. On the other hand, if you wait too long to engage a factory and already have a fully detailed design, it might not be able to manufacture your product in the way you want, or there might be higher costs associated with features than you thought.

Here are some terms you will encounter as you go through different stages of manufacturing:

Design for X (DFX)
: The process of making sure your final prototype can be be manufactured efficiently and within budget. This process comprises design for manufacture (DFM), design for assembly (DFA), design for test (DFT), and design for cost (DFC).

Engineering verification test (EVT)
: The first gate in manufacturing. At this stage you are normally hand-assembling parts from tooling to check engineering function.

Design verification test (DVT)
: The second major stage in building a pilot. You should have all tooled parts going down a custom assembly line at this point.

Production verification test (PVT)
: The final stage in production ramp before full-on mass production (MP). At this point, you are mainly focusing on the production line itself, not the product. Most startups will ship product coming off the line, since it is almost perfect.

Standard operating procedure (SOP)
   Detailed instructions of a manufacturing process spelling out each individual step. Sometimes referred to as MWI (manufacturing work instructions).

Approved vendor list (AVL)
   A list of vendors to a factory or startup that have gone through the appropriate vetting, audits, and screenings to be a trusted source.

Computer-aided design (CAD)
   The design files documenting ME and EE parts. ME CAD is normally transferred through STEP or IGES 3D files or traditional 2D drawings for critical parts. EE CAD is transferred through Gerber files and drill files for boards.

Orion Labs, formerly OnBeep, is a great example of a hardware startup that approached manufacturers early enough to get input but was still far enough along to explain to manufacturing partners what it was making. "Ramping Up Manufacturing at Orion: A Case Study" on page 115 highlights some of their process.

Ramping Up Manufacturing at Orion: A Case Study

Andy Sherman and Star Simpson have both led Production Operations at Orion Labs (*https://www.orionlabs.co*), formerly OnBeep. Star ran production at Orion as it began ramping up manufacturing of its Onyx product. Andy came on board during this ramp, and Star transitioned to R&D. Onyx is a wearable group communication platform with a button, a microphone, and a Bluetooth connection. Orion engaged PCH Access (*http://pchaccess.com/*) to help with its first production run in China.

When selecting a partner for its first production run, Star recalls, "We were looking for the ability to scale." She emphasizes that a startup should "look at what projects manufacturers have worked on and that informed their experience," since you wouldn't want someone who has never worked in your industry and would be unfamiliar with your processes. Orion ended up choosing PCH Access to help with its first production run. Having a partner helped alleviate the need for a large team at Orion to get through the first run.

Andy describes the key pivotal role in managing a partner as Orion's "program manager, someone who works with either the contract manufacturer or the other team, on the ground overseas at your manufacturing site to some capacity." Orion also needed technical staff in "electrical engineering support on the ground, mechanical engineering support on the ground, quality engineers so this way we can define quality and test plans." These kinds of technical resources are important to keep in-house so you can control the technical specifications of your product—since only your startup knows what makes your product truly unique.

As far as partner support goes, Andy describes, "On the PCH side, we had a program manager that managed the team over at PCH. We also worked very closely with an electrical engineer on that team, with a sourcing procurement expert on that team, and a couple of quality engineers as well." Having these resources available through a partner helped Orion keep its technical team lean, while helping production move faster by having local experts on the ground.

Moving from prototype to production also required changes in the design of its product. Star describes the changes necessary in moving from prototyping to manufacturing as "redesign of a board that allows you to test that in an automated way," in the sense that not only does the board still need to function, but it now also has to be repeatedly testable. They learned specifics related to their part section, such as to "include LED variation within our devices so we had some batches that were interesting." According to Andy, "We just weren't very experienced with dealing with variations in bin codes and how we need to approach that from a manufacturing and SMT perspective, and make sure that the same bin code is found across each of the LEDs within any individual device."

Orion went through two DVT cycles on Onyx to make sure all of these issues were smoothed out. Andy emphasizes the importance of improving a line's efficiency once you get to the PVT stage. One way to go about this is to "do cycle time calculations on each stage and then see where we have bottlenecks."

Andy reminds entrepreneurs getting into the hardware field that "in manufacturing, probably more so than many other disciplines, there's no day off," as you will have to deal with these kinds of issues that arise day to day on the line. Overall, Star and Andy emphasize the continual progress made throughout manufacturing and the importance of adapting your team and product from prototyping to manufacturing.

Approaching manufacturing, it is important to change your design to accommodate the necessities of mass production. This is lumped under the umbrella term DFX, encompassing *design for manufacture* (DFM), *design for assembly* (DFA), etc. This can involve minor design changes such as adding radii or nominal dimension changes for standard tools, or larger design overhauls, such as eliminating unnecessary tooling complexity by moving features to other parts or even eliminating parts altogether if their functionality can be replaced.

Another important consideration for your design is the kind of environments and situations your product will need to withstand. This part of designing is normally called *ruggedization*. The most common consideration here is the *ingress protection* (IP) rating of the product, which reflects how much water and dust a product is susceptible to through a two-digit coding system going up into the 60s, with the highest rating technically being IP 6K9K.

Other important ruggedization considerations include the shock and vibrations that the product will need to endure and any specific material or chemicals that the product will be exposed to, especially during normal cleaning.

Once you engage with a CM, it will want you to submit your design files so it can give you a quote. This back-and-forth process is known as a request for quote (RFQ), and it's important to be prepared with any documentation or information that the vendor will need in order to expedite the process. A typical RFQ will include the technical design files for whatever you are looking to make as well as accompanying documentation setting expectations for process, tolerance, quantity, timeline, and any other requirements. If you are working with a full, turnkey CM, it will want everything, including your ME CAD, EE CAD, product requirements document (PRD), BOM, and possibly even your business plan or financial information if you are starting a larger relationship with that CM.

The *bill of materials* (BOM) is essentially a list of everything that goes into your product with the goal of capturing the *cost of goods sold* (COGS), the cost it actually takes to produce your product. The BOM should include the basics of your ME and EE components (PCBs, board components, mechanical enclosures, fasteners, etc.), as well as the packaging and all of the accessories you will be including in the box. The COGS is the bottom-line cost you will be paying per unit, including all assembly, finishing, packout, taxes, tariffs, and markups from any partners. Your

BOM and associated COGS are important to start compiling early in your project to get quotes and understand what components are driving costs, but you realistically won't have a final cost until you are fully in mass production.

Another document that will become important at this point is your *product requirements document* (PRD). Typically, this captures the specifications and tests that your product will need to pass. Larger companies will start with a higher-level *market requirements document* (MRD) from the marketing team that will contribute to development of the PRD, but the PRD is often the first formal product documentation a startup puts together. It can be helpful to start putting together your PRD early in your development process, so that your engineering team has product goals to shoot toward as well as a communal document to capture qualitative considerations that can't be captured in CAD or a BOM. Your PRD can be a good catchall document during development, to capture the intent of your team and make sure everyone is on the same page about what your product is actually supposed to do and how it will perform.

Keep in mind that quotes are free but take time and resources on both sides to produce; make sure to get a few quotes, but don't waste too much time on this initial process. You will want to save this time to invest in building a relationship with the CMs and factories that you end up choosing.

Once you've released your design and issued the initial purchase orders for production, you'll need to pay for any major deviations that you would like or that the design requires. This is why it's important at some stage to put a clear freeze on your design and officially release the final design. Once this official release happens, you will need to capture changes formally through numbered *change orders* (often abbreviated CO, or ECO for *engineering change order*). Change orders often have a fee associated with them, unless it's simply a part number or documentation change.

Another important part of building this relationship is to communicate clearly with your factory, especially if any changes arise during production. Andrew "bunnie" Huang (see "Boutique Manufacturing Projects in China: A Case Study" on page 124) reminds startups that "you are not going to get your design right the first time. You are going to do some modification, so this is where the ECO, the engineering change order, comes in." bunnie once issued an ECO to a factory right before Chinese

New Year, a time of year when factories traditionally shut down for a few weeks and experience much worker turnover. The factory ended up not implementing the change order. But according to bunnie, "They went ahead and actually reworked all 200 boards by hand, free of charge, because I had this fully documented, and they readily acknowledged it was their fault for not implementing the ECO."

Another part of manufacturing that startups don't often account for is the testing of jigs and processes for their products. bunnie recommends startups "account for as much time, or more, for the design of the test jig as for the design of the core product itself." He describes test jigs as "another product," in the sense that the complexity will rival that of your original product.

When designing a test plan, it's often too easy to miss the forest for the trees. To prevent that, bunnie suggests that, in addition to working from the engineering specs, you "also visit the marketing bullet points and ask, 'What are you promising your customers?' Probably everything that you promised the customer you should explicitly test." It's also important to consider this early in the process, so that you include test points in your design and don't market claims you might not be able to guarantee through a production test.

For complex products, it's important to break down your test plan into several stages. Test the riskiest parts of your assembly before putting them into higher-value subassemblies. It's much cheaper to throw away a small part earlier in the process than wait until the full end-of-line test on the product and throw away or rework a whole assembly. Finally, tests take time, which costs money, so it's important to test just enough to be confident in your product without wasting money on redundant tests. This testing of individual units on the production line is separate from design validation and the external laboratory certifications discussed in "Certification" on page 134.

Once you start your tooling and pilot production, you will start to hit many terms and milestones. The stages of EVT (engineering verification [sometimes seen as validation] test), DVT (design verification test), and PVT (production verification test) differ slightly depending on your factory or CM, but generally, they represent increasing progress through pilot production into full-scale mass production. EVT and DVT are when you will cut tooling and set up the initial pilot production lines. By the time most projects get to the PVT phase, products are good enough to be shipped to your first customers, especially for early startups eager to get something out to crowdfunding backers.

It's important to protect your intellectual property throughout the manufacturing process. If you are concerned about sensitive IP, you should get nondisclosure agreements (NDAs) signed during the RFQ process. One strategy for this problem, especially overseas, is to purposefully split up parts of your supply chain among more factories than necessary, making sure that no one vendor has access to your entire design intent. Unfortunately, this strategy is difficult for a startup, because you will likely need to trust one or two main factories for your most important components, and especially for final assembly.

When you have ramped up the line, the next step is generally to maintain production levels and concentrate on increasing yield rates and quality concerns. This phase is typically referred to as *sustaining engineering*. It's important for startups to understand when they've hit this step, so that they can dedicate resources to the important task of maintaining production while also starting to concentrate R&D efforts toward the next product or brand line. Large companies have entire separate groups for this type of engineering, but startups often need to split the same founder between these tasks initially.

Where to Manufacture?

The decision of where in the world to set up manufacturing is important to establish early in your development process. While scale is an important indicator here, there is no "magic number" to decide when to manufacture locally, domestically, or internationally. It's also important to consider the complexity of your product. If your first product involves some sort of new technology, process, sensor, etc., you will likely need a specialized factory or vendor.

Three major manufacturing locations for US-based companies are China, Mexico, and locally in the US. This section explores the trade-offs among all of these locations. Mexico can be a popular manufacturing choice for US-based companies, because it is located closer than China, has an easier language barrier to overcome, and offers cheaper labor than the US. This can come with some drawbacks as well, as Dan Goldwater explores in "Moving Supply Chains: A Case Study" on page 121.

Moving Supply Chains: A Case Study

MonkeyLectric produces a bike light system (called the Monkey Light) that makes riders safer during nighttime rides, while simultaneously turning their wheels into digital art. The company launched with a bootstrapped first-generation kit that it made in a garage and sold on Amazon.com. Once it became clear that there was quite a bit of demand, Dan began the process of finding a factory.

MonkeyLectric considered the US, Mexico, and China. As new hardware entrepreneurs, they had a bit of trouble finding a factory in the beginning, relying primarily on Google for leads. They chose Mexico because costs were lower than in the US and it was closer than China.

"When you're small and you're making something and there's any complexity at all, you'll need to have someone there at the factory," Dan says. "If you're a small startup, that's a pretty big cost. Depending on your staff situation, it may be too difficult to send someone to China for six months. It was for us."

A typical run size was 2,000 units. The factory in Mexico would bulk-ship 500 units at a time, typically fulfilling the order within weeks. Although the quality was good and the workers were skilled and efficient, Dan felt increasingly uncomfortable with certain aspects of the factory environment and the management in Mexico. He remembers, "There seemed to be a lot of corruption and bad business practices happening in the town, and we didn't want to be putting our manufacturing dollars toward supporting that."

With three years of experience under their belt, the MonkeyLectric team decided to rethink their manufacturing process as they launched a second-generation Monkey Light. They held a Kickstarter raise to gauge interest and validate the market prior to committing the funds to the new design. When it was successful, they moved their manufacturing to Berkeley, California, and Shenzhen, China.

The current Monkey Light kit consists of two primary parts: a rubber-encased LED board, which generates the light show, and a battery holder that mounts to the hub. Each piece had its own challenges. The LED board was going to be exposed to the wear and tear of biking and to the elements, so protecting it while preserving the lightweight design was paramount. The battery holder was inspired by a camping headlight and involved a lot of small parts. Producing it turned out to be a logistical challenge. Dan notes:

> There are several pieces that go into that component—small metal springs, connectors, a rubber piece—and five or six factories make the various parts. The ecosystem for making something like it in the US is breaking down, both in terms of tracking down factories making the various parts, and finding people with the skills to put it together. It was easier to get it done in China; we only had to deal with the factory that made the final product, since their people knew where to get the subcomponents. It was also less expensive.

MonkeyLectric decided to produce the battery holder in China. Alibaba proved remarkably helpful for finding a factory capable of producing the specialized design. Dan searched the site for camping headlights, reached out to the factories that listed them, and discussed his desired modifications. He found that most producers had a much larger set of products than the ones they'd listed on Alibaba and were willing to work with him on his idea. (Dan wrote a helpful and detailed account of that process in a post on Instructables (*http://bit.ly/buy_frm_china*).

Although they looked into manufacturing the coated circuit board in China as well, the team ultimately couldn't find a factory producing anything quite like it. So they set about creating their own production line in Berkeley, California. The boards themselves are made by a contract manufacturer in Fremont, CA—no need to reinvent the wheel. Once they arrive, MonkeyLectric's team applies the rubber coating and handles quality control, testing, and packaging. The small team is capable of producing a daily batch of 400–700 units, which enables them to maintain enough stock on hand to be able to ship same-day:

> Manufacturing locally lets us rapidly iterate. We can test things out—adding glitter to the rubberized coating, trying new designs—in a way that isn't possible when you're dealing with an overseas team. It isn't always true that manufacturing overseas is necessary to be able to price a product competitively.

While overseas manufacturing can reduce labor costs, it's important for hardware entrepreneurs to consider all aspects of production—feasibility, ease of iteration, quality—when choosing the best location for their startup to set up shop.

Tesla Motors had a uniquely challenging supply chain, since it chose to take on the automotive industry, which meant that it likely needed to be near the auto industry's existing supply chain in the Midwest. Dave Lyons (see "Prototyping, and Explaining Prototypes, at IDEO: A Case Study" on page 80) was at Tesla when it was approaching the challenge of where to set up its supply chain. Dave says Tesla was originally looking at China because of the cost, as well as the timing of many consumer-electronics brands moving manufacturing to China. What it found was that partners in Asia "hadn't had the same shared experience with respect to how to get things done that weren't completely understood yet." In other words, they were good at scale, but not necessarily good at implementing new technologies.

Tesla next spent its resources looking into the entrenched automotive supply chains of the Midwest, which have traditionally relied on a multi-tiered supplier system in whic the OEMs (e.g., GM and Ford) handle all of the branding and marketing for the cars but only the highest-level engineering. The detailed engineering of subsystems and components is the responsibility of what are classified as Tier 1 suppliers (e.g., Bosch, Denso, Behr) that supply products directly to either the OEMs or Tier 2 or 3 suppliers more levels down. This system of responsibility has worked for Detroit for decades. According to Dave Lyons, the problem with innovating new technologies in that system was that Tesla "had trouble figuring out in a regular 2008 car who actually did any of the actual engineering." This made innovation hard when so many hands touched a design and none had full responsibility for it.

After learning lessons from working with China and Detroit, Tesla ended up with a rather complicated supply chain for its first cars, sourcing the bodies from Lotus in the UK and parts from all over, so it was important to it to do the final assembly locally in Menlo Park, California, in order to keep the manufacturing close to the original design engineers and collapse any potential debug cycles.

While there are many tried-and-true rules for mass production in terms of processes and supply chain, there are also many new trade-offs for entrepreneurs to consider, as bunnie Huang addresses in "Boutique Manufacturing Projects in China: A Case Study" on page 124.

Boutique Manufacturing Projects in China: A Case Study

Andrew "bunnie" Huang has been hacking on hardware since his family first got an Apple II. He went on to study electrical engineering at MIT through to his PhD. He gained notoriety on the Internet for being the first person to hack the original Xbox. He published his findings in the book Hacking the Xbox (*http://hackingthexbox.com*), a great introduction to reverse engineering consumer electronics. His first exposure to mass manufacturing was as the hardware lead on the Chumby, an embedded computer that was one of the first connected, open source devices.

bunnie now lives in Singapore and continues to work on projects related to manufacturing, doing most of his manufacturing out of China. He says that he consciously "moved from the United States to Singapore because I wanted to be close to China. I came to the ecosystem because it matters to me that much. I want to access this particular ecosystem. If you were to pick a locale, this is the locale you should pick."

That doesn't mean that he thinks all hardware entrepreneurs should immediately move to Asia, but rather:

> *If you're located in Boston, find a manufacturer in Boston. If you're in Texas, find someone in Texas. And not just in Texas, in your city in Texas. There's a lot of friction in going far away, right? Loss of communication, inefficiency of process implementation, cost of logistics, soft cost in terms of a plane ticket and your time.*

bunnie encourages entrepreneurs to work out new processes and product lines locally before scaling remotely.

bunnie says that there are great benefits to approaching the manufacturing landscape from a different perspective, in the sense that most entrepreneurs have "a temptation to say 'Oh I'll just compete on price' since you're in China, but then your margins are really thin and you're constantly cash-starved. It's really easy to fall upside down in this situation, at which point you have to raise venture capital or you end up not meeting promises."

Instead of competing in a race to the bottom solely based on cost, bunnie likes to approach the Chinese manufacturing ecosystem differently by seeing what value he can add to products through unique approaches to processes and supply chains. Two great examples of this are two of his recent projects: the Novena laptop and Chibitronics circuit stickers.

This approach of embracing cutting-edge processes is exemplified by the Chibitronics (*http://chibitronics.com/*) project, in which bunnie, along with Jie Qi from the MIT Media Lab, developed flexible electronics stickers that can be adhered anywhere, just like conventional paper stickers. bunnie remembers:

> I didn't want to get in a situation where I'm essentially trying to be the master of sourcing cheap LEDs, so we have to add some value beyond low cost, which is the peel-and-stick nature of the circuits. So we focused a lot on that process. I wanted to make sure we were doing something that actually didn't exist in a factory anywhere else. We built, from the bottom up, a line to manufacture that particular process. That's a unique value in itself.

By examining the process at every stage, even going so far as measuring raw material stock in factories to calculate the most efficient 2D tessellation strategy for wastage, bunnie was able to add value to the product that others would not be able quickly replicate, in China or elsewhere. This is important, because it also shows how having a close relationship with factories can facilitate a depth of information that can help you build value around your partner's processes.

bunnie says that this is all part of the changing landscape of manufacturing processes, which is indeed shifting, but not as rapidly as public perception might indicate. He posits:

> The most important thing to be aware of is that the equation is still basically correct, but the constants are shifting. The upfront cost of injection molding is going down; the per-unit cost of CNC is going down. 3D printing's quality is getting there eventually, but just one big problem of many that 3D printing has as a production process is the part quality doesn't have the finish or the reliability that you expect out of a comparable injection-molded part.

That said, bunnie says there are other reasons to choose processes during this convergence, even if they're not necessarily at the same level as traditional production processes:

> One of the ways you look at this is: prototype is production and production is prototype. This whole thing about injection molding versus CNC and 3D printing—they're guidelines. They're just top-level, high-level guidelines at the end of the day. Anything's possible.

There are no rules. Just figure it out and figure out what process makes the most sense for your product. Sometimes, you can go with total 3D printing as part of the story, and that can sell more units for you because there's press-worthiness in 3D printing these days. That's a completely valid reason to go with a production process.

bunnie has become a keen observer of China's unique approach to collaboration, something he discusses in "From Gongkai to Open Source" (*http://bit.ly/from_gongkai_to_os*) on his blog. bunnie coined the term "Gongkai" to refer to a distinctly Asian approach to sharing that involves information flow among many parties in order to get things made, rather than the Western approach of ideas flowing in one direction and compensation in the other. He contrasts Gongkai to the Western approach to intellectual property, where you are worried that you'll "end up spending more money on lawyers than tooling." The open approach allows smaller teams of engineers to develop products more rapidly. bunnie uses cell phones as an example, because China's open framework is able to produce phones cheaper, faster, and with more variety than seen in Western markets. Overall, bunnie sees the Gongkai approach to IP "as evidence that a permissive IP environment spurs innovation, especially at a grass-roots level."

Another recent project bunnie has taken on with collaborator Sean "xobs" Cross, Novena (*http://bit.ly/novena_update*) is an open hardware computing platform that can be used as a desktop, laptop, or standalone board. As bunnie progressed through his Novena project, it was important to limit manufacturing processes to the project scope. Since he knew from the start it was a project aimed at hobbyists, he avoided processes that are geared toward high volumes. For example, while most laptops have four major injection-molded plastic shell parts, the Novena gets away with only one.

Other unconventional aspects of the project include "the bezel that covers the connectors, we call it the Port Farm. We wanted that to be replaceable, so that as you upgrade the motherboard you can reuse the case." Beyond traditional future-proofing, bunnie says,

The design itself is driven also largely by cost constraints. Injection molding is really expensive, and you want to minimize the number of big tools that you have. If you look at our design, there's one major tub that holds all the components; that's a single injection-

> *mold piece. If you look at a typical laptop, there's effectively four tubs, so they have four times the core tooling costs that we have.*

Making sure that your manufacturing process tracks to your scale is crucial, especially when starting out with smaller production runs for fulfilling crowdfunding campaigns.

bunnie builds on that conclusion by saying that the design language of the product should match the technological philosophy:

> *I chose these socket cap screws that stick out. They're really very distinctive and out there. They're very aggressive in terms of the techy look. That's very intentional. I want you to see the screws, because you're going to deal with those screws. That is going to be your experience in this laptop. That's part of the language of the design.*

It's important to be honest to the aesthetic of your user group, whether that design language ends up being techy, rugged, elegant, or something else that connects with your primary users.

bunnie also stresses the importance of choosing your factory partners and maintaining clear communication and a good relationship with them:

> *I consider them almost at the same level as VCs in the hierarchy of partnerships. If you can negotiate with the factory and they can treat you like a real partner, they will effectively use their balance book and their buying power to go ahead and reduce the amount of capital you will have to raise to go ahead and get your product out there.*

These kinds of benefits are often overlooked by people coming from software or other nonproduct backgrounds. The working capital required to get parts into a factory and products out can be huge enough to sink a startup if it doesn't have a fast channel to customers and is on bad financial terms with factories, distributors, or retailers.

This is often best established through a good relationship with the factory owner. bunnie says, "If you go to a factory and the boss does not come out and at least poke his head in or say hi to you, it's a bad sign, because you are dealing with a mid-tier sales rep." Once a factory boss is convinced your project is a winner, you will find it gets priority handling and you're working with an A-team staff, something no amount of money can buy.

> Overall, bunnie stresses the importance of the relationship you build with your partner factories. This can be facilitated through clear communication and respect, and it will benefit your startup in the long term through access to new capabilities, clearer communication from their side, and even potentially better payment terms and other more direct benefits.

Supply Chain Management

You will need to set up a supply chain of multiple factories. For example, the plastics for an enclosure are likely injection-molded at a factory dedicated to plastics. The electronics PCBAs would be produced at a completely different factory. Assembly usually happens at an electronics factory, since there are cleaner environments and more testing equipment available. Packaging, batteries, specific sensors, and specialized mechanical components would come from even more factories. The product is finally put into the packaging, along with any paperwork and accessories such as charging cords, at a packout or fulfillment center. From here, it is either shipped directly to the consumer, or put in larger master cartons with other boxes if it is being shipped to distributors for retail. Even this is a simplified version of typically multitiered and complex supply chains. A supply chain can go as far back as the materials, or even raw ores, that are mined to start producing parts.

Another important topic to consider when examining your supply chain is the entire life cycle of your product. How will consumers recycle, repair, or dispose of your product when they're done using it? This is especially important to consider for objects like lithium ion batteries, which can't be thrown away in typical garbage streams. You can help this issue by providing return and disposal instructions in your manual and online. You will definitely need to include provisions for this if your product involves disposables in your business model that you need to recapture.

You will also want to think through how a user will repair your product, especially for simple operations such as battery replacement and replacing parts that might be prone to breaking. This can allow you to design these repair and replacement experiences to be less painful for users, and also get ahead of anything the public is bound to figure out about your product when iFixit does a teardown (see "Teardowns" on page 86).

Complicated products such as cell phones can have dozens of factories involved in production through many tiers. These kinds of supply chains can get so complicated that buyers in a large company will typically source components from multiple vendors or factories just to ensure that if a vendor can't meet commitments, it doesn't impact the supply chain and production can continue. This approach is a common way to mitigate supply chain risk and give buyers bargaining power on price, since both sides know they could get the same part elsewhere.

You can also rely on a *supply chain management* (SCM) company to set up and/or run this process of finding factories for you, but you will have to pay it a percentage of your BOM and/or an upfront fee. Alibaba (*http://www.alibaba.com/*) has become a popular website for finding factories and suppliers in China, but it can be a mix of interactions; some listings are actual factories, but many are distributors, holding companies, or other middlemen with little knowledge of the actual product.

Importing from Foreign Manufacturers

If you manufacture in a foreign country and then bring the goods into your home country, you are an *importer* and subject to a series of reporting requirements and tariffs. For the purposes of this section, we're assuming that you're importing to the US.

Importation can be a challenging process. The process of bringing goods from an overseas factory into the US generally happens in one of two ways: air freight or ocean freight. (If you're in the US and manufacturing in Mexico, trucking is an option.) The means by which you import will dramatically affect your costs. In general, air freight from Asia costs approximately four to five times the cost of ocean freight. However, ocean freight shipments can take a full month to arrive, and your shipment might be bumped (shifted to another ship) and delayed several times. The weight and bulk (volume) of your item is an important consideration in deciding which method to use. So is the urgency of customer demand; sophisticated companies often determine a percentage of goods that they'll import by air to meet immediate demand, and then ship the rest by ocean.

Once your container has arrived at the port, it's loaded onto a truck or rail to reach the distribution center or warehouse. This movement of the container from one type of transportation to another is called *intermodal freight transport*. Pallets arriving by air are also loaded onto trucks. To ensure a smooth transition from one mode to the next, you'll likely be employing the services of a freight forwarder. This is particularly true while you're a small company; only large shippers contract directly with major ocean carriers.

Freight forwarders will book passage for your goods. An ocean container may be tough to fill up when you're first shipping; the smallest option is a 20′ box. *Consolidators* let you ship LCL ("less than container load"). They aggregate your shipment with the shipments of others to secure a more favorable rate. Freight forwarders can also help navigate intermodal transitions (e.g., ship-to-truck or ship-to-rail), including getting the goods from the factory to the port of departure, or from the port of arrival to the distribution center. Their fees may include insurance for your shipment while it moves through transit (if not, purchase cargo insurance), warehousing costs for temporary storage between stops, arrival agent (*http://bit.ly/arrival_agents*) fees, and more. They are middlemen, and there is little in the way of price transparency in the industry; you'll want to call several companies to identify the best partner. You can also ask your factory if it will help you book transport and compare the rates you receive with its. Some factories will incorporate shipping fees as part of a per-unit cost.

One of the challenges of working with a forwarder is that it's often difficult to understand exactly what you're being charged. What part of the rate is the truck or the ocean passage, and what part is going to the freight broker? The process of comparing quotes can be difficult and time-consuming. This is further compounded by the fact that many forwarders communicate only by phone and fax. A few startups are attempting to bring price transparency in the freight space. One is Haven (*http://haveninc.com*), which enables customers to use its web platform to obtain quotes directly from carriers and bid for capacity based on trending prices.

One common term you will come upon when looking at importing and shipping is *free on board* (FOB), which designates which port your product will actually be shipped to. There is a big difference in price if your logistics partner is quoting FOB Hong Kong versus FOB Long Beach. This is defined even further by the latest standard Incoterms (*http://en.wikipedia.org/wiki/Incoterms*) established in 2010.

What to Look for During Manufacturing

This section covers the most important areas for each technical team to address during manufacturing. The roles on your team will be shifting as you move beyond creatively making one prototype into a more rigorous mindset of reproducing the final design over and over again.

By this point in your development, the project should have been handed off clearly from your design team or partner to the engineering group. If you have industrial designers still involved this late in the process, they will likely be technically unequipped to converse with manufacturing engineers, and might even be suggesting design-level changes that would have an expensive ripple effect to implement. It's better to let the engineering or operations team handle things from here on out.

Mechanical engineers are usually most concerned with the tooling that is developed for your custom mechanical parts. *Tooling* is a broad term that refers to any of the molds, jigs, and fixtures that will be fabricated purely to produce your product but not make it into your final product.

The most common (and expensive) type of tooling for hardware startups is often steel molds for injection-molding plastic parts. Injection-molding tooling is important to address early in your release cycle, because lead times can be 10+ weeks for traditional steel tooling capable of making hundreds of thousands of parts. This tooling can cost $10,000–$100,000, depending on the size, complexity, and material of your parts.

For a startup making fewer parts in a first run, you might want to consider *soft tooling* in aluminum to move faster. 3D printing of molds themselves (typically through a direct metal laser sintering, or DMLS, process) is also currently an experimental process that has been evolving rapidly. *Jigs* and *fixtures* also refer to dedicated tooling, but a jig or fixture is usually more common for alignment, assembly, or letting parts cure in a stable position than molding an entirely custom part.

Mechanical engineers will commonly encounter iteration of the tooling to get closer and closer to the design intent. Commonly, the first parts off an injection-molded tool are referred to as "T0" parts. They will not have the texture of the final part and are expected to not be fully correct, but they provide a good gut check for how much iteration will be necessary.

T1 parts follow, after an iteration has been made on the tool, and this is often the stage in which you will add texture to the tool. More iterations would be subsequently named T2, T3, and so on, but if you go through too many iterations here, you might end up being better off starting the tooling process all over again with a fresh piece of steel. This can be costly, which is why it's important not to wait too long to engage manufacturing experts to make sure you are designing parts that can be tooled.

At some point, when you are satisfied with the parts coming off the tool or any process, you will want to go through a formal *first article inspection* (FAI), during which you measure all of the *critical-to-function* (CTF) dimensions called out by your design engineer and produce a *first article inspection report* (FAIR) documenting which dimensions are in spec and which are not. This is important for your quality team to look back at later and understand how the first parts came off so that they can assess issues later to see if the tool has "drifted" out of spec through repeated use.

This can also be a good time to pull out one *golden sample* for later reference. A golden sample is a part or assembly that is completely within spec and deemed essentially "perfect." Golden samples are especially helpful for areas such as color matching and other qualitative properties that need to be evaluated.

During your development process, it is helpful to consider features of your design that you can make *tool neutral*. Because steel tooling is a negative form of the part you are trying to create, *tool neutral* refers to the fact that it is much simpler to make a feature *larger* by machining the tool. Making a feature *smaller* requires welding additional material back into the tool. Buttons and other areas requiring a good-feeling mechanical fit are prime areas you will want to purposefully design too small at first, knowing you can subtract tool steel later to add plastic to each part. This process of "dialing in" the feel is common and is part of why no specification can replace the value of having one of your engineers on the line when parts are first coming off tooling.

Electrical engineers typically manage the build of the PCBAs during manufacturing, to make sure everything is functioning as desired before putting it into an enclosure. This involves overseeing fabrication of the bare PCB boards to specification; correctly assembling components onto that PCB through a pick-and-place machine or manual insertion; soldering these together through a reflow oven and/or a wave solder process; and finally, inspecting the board before further assembly.

Firmware engineers will work closely with the electrical engineers to ensure that the line is being designed to test the limits of the board. Often, they will need to create custom test firmware for the device as it goes down the line to push the limits of a wireless antenna or run self-tests of specific components, such as accelerometers—all without burdening the final firmware with the overhead of this code. The device will then be flashed with its final firmware before going through the final, integrated end-of-line test.

Something that will become increasingly important when you enter manufacturing is *quality*, which refers to the fitness of a part or product for its purpose. This is usually broken into two main categories: *quality assurance* (QA) and *quality control* (QC).

QA refers to the overall technique of improving your processes and production line. QC is the process of actually measuring parameters to see if parts are in spec. This often happens for parts coming into a factory, to ensure that you are starting with known good parts and not wasting time assemblying broken chips into an otherwise functioning product. This is referred to as *incoming quality control*, and you will have other QC touchpoints along the production line as well.

Quality processes will be crucial for making sure you have a high *yield rate*, which refers to the number of parts or products you manufacture that are within spec and can be used. Yield rates vary, depending on the process and new complexities you are adding. You hope to eventually get them into the desired 95–99 percent range, but you should realize that it will be much lower than this during your first run. 70–90 percent yield can be common for first runs, and some more complicated processes can create yield rates under 50 percent.

It's important to consider that this is also a function of your spec. That is, if you loosen your standards for what is in spec, a 90 percent yield rate could become 100 percent. This is tempting to do in the late stages of manufacturing, when you feel pressure to get the product out to customers, but you should always look back to your original spec and hold your product to a standard you are proud of.

It can also be interesting to consider alternative manufacturing processes and materials during manufacturing, because they can potentially save you time or money, and might also provide a marketing angle if you want to brand yourself as a sustainable or environmentally friendly company. It's important for new processes to always be innovating, but to avoid risk, you should also look toward processes that have been shipped in a product by someone else. You also want to be careful not to include sustainable practices purely for bragging, because this can backfire if your company is suspected of as *greenwashing*.

Realistically, resource-strapped startups might have only one or two people wearing all of these different hats during manufacturing, especially if manufacturing is happening overseas or far away from your office. It can be helpful when manufacturing abroad to have a constant rotation of employees going over to check on production, especially at key times such as initial line bring-up and process changeovers.

Certification

Before you begin mass production, you will need to go through the process of certifying your product with different regulatory agencies, depending on the geographical areas you are launching in and your product type.

Two very common certification groups you will need to deal with if launching in the United States are the Federal Communications Commission (FCC) and Underwriters Laboratories (UL). The FCC regulates anything that is transmitting information wirelessly, which many hardware products will do through Bluetooth, WiFi, or GSM. UL certification covers many different product types and industries, including information technology, medical, power and control, appliances, life safety, and security, to name a few. They also do FCC testing and CE for European countries, among many other certifications.

UL has a few pointers for hardware entrepreneurs going through the certification process. Although some products might not be mandatory to certify, startups will also voluntarily certify other products in order to protect their own liability from safety issues. UL recommends that, when working with a startup on a new or existing product, knowing their marketing plan is important, because it goes hand in hand with your compliance plan.

This is important, because your compliance is based both on what claims you are marketing for your product, as well as what geographical areas you are launching in and shipping your product to. Knowing all of this up front helps ensure startups are taking the correct approach to regulatory compliance, as it can balloon into an expensive process. It could go anywhere from $1,000 to $100,000 or even more, depending on the product, types of certifications, and countries in which you intend to sell.

That's why it's important for startups to consider certifications during their design and prototyping phase, in order to reduce cost and rework, while increasing speed to market by avoiding surprises through the product life cycle.

It can be helpful to work with a local test lab early in your process. They can help you identify which tests you will need to pass by walking through your intended use case with you. This can affect your design, so it can be helpful to get feedback from a test lab earlier rather than later. They can help you as well by doing prescans on final prototypes that can help identify noncompliant areas or anything you can tweak before the final certification test. A friendly test lab can save you lots of time in researching specs, as well as proper design guidelines, and make your certification go smoother by helping with preliminary investigations and prescans.

Ben Corrado, cofounder of engineering consultancy Rigado, has some advice for startups going through regulatory certification. If you are experiencing *electromagnetic interference* (EMI) issues, Ben says that some products will end up "spraying the inside of your whole plastic enclosure so that it's metalized," but, he says, you can often blunt this by "adding additional layers to the board and sandwiching it all to copper."

Ben adds that "it is a very iterative process. We 3D printed a device and we wrapped the whole thing in foil tape and we slowly peeled back the foil tape." Iterating different enclosures like this can often be a much more time- and cost-effective approach to EMI shielding than long cycles with expensive analysis software. Ben also points out an important difference when going for CE marks necessary in Europe. During the FCC's certification process, "all you're doing is radiated and conducted emissions. When you go to CE, you're doing immunity as well, so you're going to have things blasted at you."

There are also many industry-specific agencies and trade groups with standards that you should consider for your product; examples include the Society for Automotive Engineers (SAE) for automotive products, National Institute of Standards and Technology (NIST) for measurement products, or the National Sanitation Foundation (NSF) for food-related products. Apple's Made for iPhone/iPod/iPad program (MFI) is an example of a certification of compatibility with a specific company's equipment. These kinds of trade groups are more industry-specific than the previous general consumer-product certifications, but they will often have standards you will need to meet before anyone in your industry will stock your product.

Another category of certification so complicated that it could be the subject of its own book (and there are plenty of books dedicated to it) includes those provided by the Food and Drug Administration (FDA). This certification is required for any medical products and some food products, and includes specific categories and regulations, depending on how invasive a product is and what level of recommendations is being made. Most consumer products avoid many of these regulations in their first product by branding and marketing as a measurement device, not a diagnostic or medical device.

The most important part of certification is knowing early in your process what certifications you need so that you can make the correct design considerations during prototyping and work with an engaged test house during manufacturing to make sure this doesn't become a time setback in the ramp-up to production.

Packaging

Another important consideration in manufacturing is your packaging. This is important to think about before you enter production, but it shouldn't really be on your mind until after you figure out what product you're making. A package's primary function is to protect the product during shipment, distribution, and retail, but it's also important to consider the branding potential available with your packaging.

Once your product is successful enough to sell at retail outlets, your package itself becomes a great way to sell your product. Obviously, there is lots of marketing and branding to make sure people are aware of your product before they go to a store, but once they are on the aisle comparing your product to all of your competition, your package has the responsibility to inform and differentiate your product.

Before designing your retail package, you should visit the retail stores you envision stocking your product and find the areas in which you expect your product to be sold. You can then consider how your product's packaging might fit in there. For example, is everything sitting on shelves or hanging from tags? If all the boxes in the area are blue, should yours be blue to fit in or red to differentiate? It's also important to make sure your packaging is large enough to incorporate all of the branding, logos, and messaging you need to differentiate without being unnecessarily large, so you don't waste material, shipping costs, and retail space on the package.

Beyond retail shelves, it's also important to consider how your products get packed during shipment. Usually, packages are placed in a larger box called a *master carton*, and then attached to a pallet. It's important to consider dimensions, so that you can maximize the number of boxes in a master carton and the number of master cartons per pallet. This can often be the fundamental trade-off to decide while designing your package: do you increase size and maximize branding space on the package, or do you minimize the package and optimize for cost? The right answer for your product will almost always be in the middle of these extremes.

oDuring packaging design, it's important to note that a *packaging designer* who deals with the structural form of the box will likely not be the *graphic designer* who chose the colors and designed the graphic assets of the company (logos, fonts, etc.), which were probably made for your website and launched long before you began considering packaging for the product. Separating these two disciplines can help you not only go faster, but also focus on the importance each discipline brings to the package.

Out-of-box-experience (OOBE) is something that has become more and more important for consumer products. Apple has set a high standard here with the iPhone and iPad; that has driven up customer awareness of the package and increased the cost companies are willing to spend on packaging. Customers should have a pleasant experience opening your box, while also learning more about your product and how to use it while they unpack. There should be a logical flow in how the box is unpacked. You almost always want to start with the featured product for emphasis and bury cords, manuals, and accessories another layer down in the packaging. The package can also sometimes be repurposed for the product so that it isn't wasted. For example, it could become a charging station, mounting stand, or other natural accessory for the product.

Before you jump to a full retail package with 10 colors, foldout panels, and a viewing window, you might consider a *white box* approach for your first product. This means that the box has only the base color (likely white or natural cardboard) and printing in one color (normally black) on one or two sides. This minimizes the cost of your package, because this version will likely never be on a retail shelf where the box might need to drive sales. This is also effective once your product has crossed the threshold to online retail sales at Amazon, as fancy graphics and multiple colors are wasted on a customer who is making a decision based more on the online reviews and other added information, not a package in her hand.

Sustaining Manufacturing

Sustaining manufacturing is also an important subject to consider. While your first challenge is getting a pilot production line up and running, the burden it takes to keep that line running smoothly (especially when you scale up quantities) shouldn't be underestimated. This is typically the point in production where the ownership transitions fully from the engineering team to an operations team.

While it's important to have operations staff, either from your manufacturing partner or from your own team, involved in the process well before this, it's also important to take a leap at some point and fully transition ownership to the operations team. They will be better equipped to deal with the types of process-specific issues that arise, as those issues will likely be caused more by tool wear and process drift than issues with the original design. This transition also allows the design engineers to go back to the drawing board and start working on the next product or new brand line for the company.

Now that you've gotten through prototyping, successfully manufactured your product at scale, and have your production line humming, it's time to look to your future products and start developing your next product.

CHAPTER 7

Acceleration

THE PREVIOUS CHAPTERS IN THIS BOOK HAVE FOCUSED ON VALIDATING an idea, gathering potential early adopters, and creating both functional and form prototypes. Now we're ready to begin talking about how to turn your fleshed-out idea into a business.

For almost 10 years, incubator and accelerator programs have helped software companies get from idea stage to market. Incubators and accelerators are slightly different entities: *incubators* typically recruit talented people, then form companies and spin them out. They are often run within a larger company, which might pull from an internal talent pool. This model is most common in sectors in which new companies benefit from access to the larger company's infrastructure or labs, such as biotech and clean tech.

Accelerators accept preformed teams, many of which have been working on a specific idea for a minimum of a few months, often longer. The goal is to help the team scale more quickly by providing them with access to mentors, capital, and (in some cases) customers. A class or cohort is common, and there is often a set graduation date culminating in a "demo day" attended by investors. Companies in each class benefit from the feedback of the other entrepreneurs going through the program alongside them.

Due to the current popularity of entrepreneurship, a steady stream of new incubators and accelerators are popping up all over the world. According to the Seed-DB website (*http://www.seed-db.com/accelerators#*), at the time of this writing, the count was 232 programs worldwide, which have accelerated a total of 4,507 companies.

A majority of these accelerators, such as Y Combinator or 500 Startups, are *broad-spectrum*: they accept companies that span many sectors and industries. Others are focused on advancing a particular mission (e.g., nonprofit startups) or promoting entrepreneurship in a specific region or city. Vertical-focused incubators, such as Rock Health, tailor their programs to a particular sector and offer extensive expertise and connections within that sector. Some, such as TechStars, change their focus on a per-class basis, by partnering with companies that bring a particular expertise to a specific class (e.g., TechStars partnered with Kaplan to form the Kaplan EdTech Accelerator, which focuses on education startups, in 2013).

The curriculums and duration of startup incubators vary, but most share certain characteristics. There is a strong emphasis on iterative development: startups set goals and have regular meetings with program advisors to discuss their projects. Mentors play an important role, sharing expertise and advice. The startups in the class typically cowork from the same space, which allows them to bounce ideas off of one another, and they often present to one another on a regular basis, sharing challenges and solutions. Speakers come in to teach classes on specific facets of hardware entrepreneurship. Programs typically culminate in an investor-focused "demo day," in which the graduating startups are introduced to venture capitalists in an effort to make fundraising easier. The accelerator experience can be extremely valuable, particularly for new founders.

While many business-building skills are applicable to both hardware and software companies, hardware founders typically face additional complexities that aren't covered in the curriculums of even the best software accelerators. There are unique challenges associated with manufacturing, supply chain, inventory management, and fulfillment, for example.

As hardware entrepreneurship has become increasingly popular, accelerators have been created to address the specific needs of young hardware businesses. They offer the same core value propositions as the software incubators (money, expert mentorship, and a network) but also have carefully curated relationships with the manufacturers and design firms needed to successfully bring a physical product to market. Even companies that have moved beyond the prototype stage can derive a lot of value from participating in a program, as you'll see in "littleBits: A Case Study" on page 141.

littleBits: A Case Study

Ayah Bdeir, founder of littleBits, originally came up with the idea of tiny, magnetically connected circuit components as a prototyping tool for the designers she worked with as a fellow at Eyebeam. She envisioned a tool that would let designers integrate light, sound, and sensors into their projects.

She put a video of her first prototype up on her website and suddenly found herself fielding orders and requests from parents, teachers, designers, and engineers all over the world. Ayah realized that she was onto something and decided it was time to turn the concept into a product. Three and a half years later, she decided to start a company.

"I started working on littleBits in 2008," Ayah says. "By the time I decided to start a company, it was 2011. I had made the product already and raised $850,000 to place an order with a factory in Fremont, California. We started selling in early December 2011, and within two weeks we had sold out."

She took littleBits to a toy fair and suddenly had more orders than she could handle. The factory in Freemont couldn't scale, and Ayah decided that she needed to look overseas. "One of my concerns throughout was that I wanted to build a solid business on solid grounds, so manufacturing for scale—and preparing a team and infrastructure that would scale—was crucial." She turned to PCH in June 2012 and became one of its Accelerator companies.

The decision to join an accelerator primarily came down to execution speed. High demand had the six-person littleBits team working around the clock on sourcing raw material, managing inventory, and navigating the challenges of manufacturing. The team was spending all of its time simply focusing on meeting existing orders, and wasn't able to devote attention to expanding sales channels and reaching new markets.

"I decided that I wanted us as a company to build expertise in the things that no one else could do, like understand how people play, nail the design, nail the experience of a modular set that comes alive," Aya says. "The other things...other people were better than us at that, so we would partner with them."

PCH became littleBits's manufacturing and supply chain partner, their eyes and ears on the ground in China, interfacing with manufacturers, suppliers, and fulfillment services. Prior to establishing their partnership, littleBits had manufactured and sold approximately 4,000 units in total. When the first products produced by PCH came out in November 2012, it sold thousands of units within a few weeks and shipped to 45 different countries.

> As an Accelerator company, littleBits got preferred rates for these services, rates that it otherwise would have been unable to command as a young company with relatively small volume compared to big-box electronics manufacturers. Incentives between PCH and littleBits are clearly aligned. As an equity stakeholder, PCH has an incentive to help littleBits grow and increase its market share. In turn, PCH also gains a window into the lean processes of a fast-growing startup and connections to the hardware startup ecosystem.
>
> Ayah feels strongly that it's extremely difficult to bring a product to market without the help of a solid support system. "PCH is an advisor and a partner," she says. "As a founder, you have to surround yourself with people who will help you; there is no point in trying to relearn what the industry has known and been doing for decades, because that takes time away from you making something new."
>
> For any founder, giving up equity is a difficult decision. It comes down to determining whether the value-add of an accelerator increases the size of the company pie by a sufficient amount to justify the dilution. "The way I see equity, you can own 100 percent of nothing or 10 percent of a billion," Ayah says. Just be sure to be selective. Finding the right accelerator partner can help a founder scale a company more efficiently and sustainably than working alone.

If you're considering the accelerator route, it's important to know what criteria make you a good candidate for admission. While each accelerator evaluates on a case-by-case basis, some are more likely to accept a particular flavor of device, or might be more willing to take on nontechnical founders or nontraditional teams. Some require that companies be at a particular stage, while others are happy to accept preprototype ideas. Let's take a look at the options.

Lemnos Labs

Based in San Francisco, Lemnos Labs (*http://lemnoslabs.com/*) was one of the first hardware accelerators to appear on the scene. Jeremy Conrad and Helen Zelman launched the program in 2011, after noticing that many hardware startups were encountering challenges with production runs, design delays, and safety recalls. They set out to define a two-pronged process to mitigate some of these problems: regular formal design reviews, and expert guidance from mentors with experience in manufacturing, logistics, reliability, standards, and testing.

In addition to teaching and mentorship, Lemnos provides working space, access to prototyping and testing facilities, introductions to contract manufacturers, suppliers, and business operations partners. The amount of time that a startup spends at Lemnos varies according to the needs of the company; some have spent six months working out of the space, others longer than a year. The Lemnos team believes that different hardware takes different amounts of time and wants their companies to be able to mature and hit their milestones without the constraints of a more traditional accelerator "class" structure.

Ultimately, the goal of the program is to build a technology or product capable of attracting seed or Series A investment. To date, Lemnos has funded over 20 companies, ranging from Blossom Coffee (a highly specialized coffee-brewing system), to the Bia women's sport watch, to Airware, an unmanned aerial vehicle (UAV) development platform.

"One thing that stands out about our portfolio is only 30 percent of our startups are consumer focused," Jeremy says. "We love highly technical and complex systems and embrace aerospace, robotics and systems engineering companies. We're also not afraid to get our hands dirty in unsexy businesses like trash, parking, and vending." Lemnos Labs invests up to $200,000 in each company and prefers to be the first money in. On average, it invests $100,000 for a 10 percent equity stake.

Here's Lemnos Labs' successful applicant profile:

Ideal founder background
Technically minded founders. No entirely nontechnical teams.

Team size
2+.

Stage
Prototype; a duct-taped model is fine, but there should be something.

Prior hardware experience?
Experience in industry or academia is looked upon favorably; most successful applicants are at least a few years out of school.

Prior funding?
Up to $500,000.

Acceptable sectors
Robotics, consumer devices (all fields), appliances, aerospace, transportation.

Application process
Email application, which should include team bios, a description of the technology, market opportunity, a business model, and a rough budget/development plan.

HAXLR8R

Pronounced "hack-celerator," HAXLR8R (*http://www.haxlr8r.com/*) is a class-style accelerator for hardware entrepreneurs. It is unique in that the program takes place in Shenzhen, China. It offers $25,000 in funding (with the option of an extra $25,000 convertible note), workshop space, and a 111-day program to help companies grow from idea to product.

Participants spend the majority of the 111 days in Shenzhen, which HAXLR8R founder Cyril Ebersweiler calls "the factory of the world." The last two weeks (and demo day) take place in San Francisco, California. Entrepreneurs spend their time in the program getting to a refined prototype and sensible business model, and gaining an understanding of the Chinese manufacturing ecosystem.

HAXLR8R places a good deal of emphasis on founders being close to factories, focusing on lean manufacturing with Chinese engineers who view speed and cost as design and business constraints. According to Cyril, "No hardware plan survives contact with a factory."

An early introduction to the factory owners and being on site to watch production workflows helps new founders develop an understanding of the complexities of manufacturing. These relationships alone would take years for an entrepreneur to develop on his own.

The curriculum covers prototyping, product design, sourcing, manufacturing, fulfillment management, and fundraising. HAXLR8R typically takes a 6–10 percent equity stake. So far, it's held five "classes" and has 50 alumni companies.

Here's HAXLR8R's successful applicant profile:

Ideal founder background
Varied; passionate and hard-working.

Team size
2–4.

Stage
: Preprototype or prototype.

Prior hardware experience?
: Not necessary.

Prior funding?
: Perfectly acceptable; just disclose it in your application.

Acceptable sectors
: Robotics, consumer devices (all fields), appliances, biohacking, gadgets/toys ("of a disruptive nature").

Application process
: Online application, including a video pitch.

Additional factors
: Looking for ideas that solve real problems or create a meaningful technological advance.

AlphaLab Gear

Another incubator run by an investment fund, AlphaLab Gear (*http://alphalab.org/gear*) is the hardware-specific offshoot of AlphaLab (*http://alphalab.org/*). This new eight-month program is a collaboration between Innovation Works, Pittsburgh's largest seed-stage investor (and one of the most active on a national scale), and Startbot, a fund focused on early-stage robotics businesses.

Teams work out of the 10,000-square-foot dedicated AlphaLab Gear facility and receive memberships to Pittsburgh's TechShop. They also have access to a workroom with engineering and prototyping technologies. The program's extensive mentorship network helps provide expertise in design, engineering, manufacturing, and sales. Teams must physically be present in Pittsburgh, PA, for the duration of the program.

AlphaLab's network of other partners offers various perks as well, including free legal and accounting services, no-cost components and development kits, engineering assistance, and industrial design. The program awards $25,000–$50,000 per team, in exchange for a 5–9 percent equity stake. The goal of the program is to get teams to the point of being ready to raise funding at the end of the eight months.

Here's AlphaLab Gear's successful applicant profile:

Ideal founder background
No set formula, provided that founder backgrounds includes experience in the areas critical to the success of their specific business. This tends to result in the program accepting primarily technical founders.

Team size
2+; while it has accepted solo founders in the past, the likelihood of success as a company is much lower.

Stage
From idea to postrevenue; anything is fine.

Prior hardware experience?
Not required.

Prior funding?
Acceptable to have raised money prior to the start of the program, but not required.

Acceptable sectors
Any company with a physical product component. This has historically included robotics, automation technology, additive and subtractive manufacturing tools, sensors, consumer products, automotive products, and health care products.

Application process
Online application process, interview.

Recently, several large supply-chain management companies have launched their own hardware incubators. With core competencies in manufacturing, logistics, and fulfillment, they offer hardware startups the opportunity to scale through strategic partnerships and access to industrial-scale facilities, in addition to equity investment.

PCH

PCH (*http://www.pchintl.com/*) designs custom manufacturing solutions for large and small companies. Since 1996, PCH has provided consumer-electronics clients services such as product engineering and development, manufacturing, kitting, distribution, and all the stages in between.

Product development is handled by PCH Lime Lab, a team of former IDEO, Apple, and Design Within Reach employees that was acquired by PCH in June 2012. The PCH team purposely does *not* do industrial design, because they feel that it's different for each brand and specific to the product and the company, and because they don't want to sell the wrong service.

PCH has established relationships with hundreds of factories, as well as its own six facilities (focused on specialized manufacturing, kitting, fulfillment, and distribution) in the Shenzhen Free Trade Zone. PCH employs over 2,500 people in China alone, including hundreds of mechanical and electrical engineers. PCH helps its clients make the most of the advantages (and avoid the difficulties) of manufacturing in China. PCH actively promotes sustainability and humane treatment of workers: it works closely with nongovernmental organizations such as workers' rights organization Little Bird, and has opened hotlines in its factories so that factory workers can report their experiences and grievances directly to top management anonymously. For more information, you can read PCH's sustainability report at PCH sustainability (*http://bit.ly/pch_sustainability*).

In 2015, PCH acquired Fab.com to add retail to its offerings.

PCH recently began to work more closely with startups via two focused accelerator programs: Highway1 and PCH Access (formerly PCH Accelerator).

HIGHWAY1

PCH's Highway1 (*http://highway1.io/*), focuses on early-stage startups. Highway1 accepted 11 companies into its Autumn 2013 inaugural class and has graduated a total of 35 companies as of December 2014. Brady Forrest, vice president of Highway1 (and coauthor of this book), says, "PCH is interested in companies that are pushing the edge of technology. We're interested in growing the ecosystem. We want to help the best entrepreneurs become great hardware companies."

Highway1's four-month program is based in San Francisco, California, with two weeks spent in Shenzhen, China, for factory and electronics-market tours. Each team receives a $50,000 cash investment, office space, mentorship, marketing and business support, and 24/7 access to a prototyping lab that includes electrical and mechanical engineering tools.

Highway1 companies also benefit from a dedicated team of PCH engineers and other subject experts to help them bring a product from concept to working prototype, and beyond. The Highway1 equity stake is negotiated according to the stage of the company, typically between 4 and 7 percent.

Here's Highway1's successful applicant profile:

Ideal founder background

An entrepreneur, risk-taker, big thinker, engineer, designer, business person, hustler, hacker, and maker. Some of those skills may be rolled into one person, but ideally they are all represented within a team.

Team size

2–4.

Stage

Prototype, past the sketch stage. Ideally, teams have produced both a functional prototype (even if it's a board and some duct tape) and a model that conveys a sense of the design.

Prior experience?

Prior hardware experience is nice to see, but not required. PCH will accept founders right out of college.

Prior funding?

Perfectly acceptable to have funding.

Acceptable sectors

Primarily focused on connected hardware and home goods. They are interested in B2B hardware as well.

Application process

Online application, including a video pitch. Many come in through a network of referrals from investors or other hardware entrepreneurs. They welcome founders who wish to get to know them early.

Additional factors

Based on its own experience as a multinational company, PCH believes that founders benefit from international diversity. It likes to draw from teams from all over the world to fill its cohorts.

PCH ACCESS

PCH's second program for startups is called PCH Access (*http://pchaccess.com/*). This program is a long-term, comprehensive, growth-focused experience designed for small companies that are looking to scale and take their product to market. Highway1 companies can "graduate" into PCH Access. PCH Access also accepts companies from other accelerators, and directly into the program.

For Access companies, PCH offers the same types of services that its Fortune 500 clients receive. Offerings include product design and engineering, manufacturing, testing, regulatory approval, production, pack-out, and delivery to customers—a complete end-to-end process to help entrepreneurs scale, save time, and avoid costly mistakes. PCH also provides a line of credit, taking payment upon sale, which is extremely valuable for young companies with limited working capital.

As hardware entrepreneurship has become increasingly popular, accelerators have been created to address the specific needs of young hardware businesses. They offer the same core value propositions as the software incubators (money, expert mentorship, and a network), but also have extensive relationships with the manufacturers and design firms needed to successfully bring a physical product to market. PCH Access builds on this model, helping entrepreneurs grow their companies. Even companies that have moved beyond the prototype stage can derive a lot of value from participating in a program like PCH Access, as you saw in "littleBits: A Case Study" on page 141.

Flextronics

Flextronics is a Fortune Global 500 supply-chain management company with factories and operations in more than 30 countries. It touches all aspects of physical product design, from design through manufacturing, distribution, and fulfillment. Its clients include Apple, Motorola, Cisco, and Microsoft.

In 2013, Flextronics announced Lab IX (*http://www.labix.io/*), its early-stage startup acceleration platform. Startups accepted into the program receive access to its engineering services for manufacturing, testing, and tooling. They have a design partnership with IDEO and a shared workspace laboratory to facilitate rapid prototyping. The Flextronics program is a rolling admissions process; there are no structured "classes."

Flextronics provides up to $500,000 in the form of a mixed package of cash and services. It also provides a credit line for additional services: up to $200,000 with no purchase orders, increasing once the company has hard orders from customers. This is in exchange for an equity take of between 3 and 20 percent, depending on the stage of the company.

Here's the Flextronics Lab IX successful applicant profile:

Team size
Flexible; teams can be as few as 3 people, as many as 20.

Stage
Existing beta or functional prototype demonstrating product capability, with solid business plan.

Prior funding
Flextronics prefers founders who have raised seed funding.

Acceptable sectors
Aerospace, automotive, homeland security, medical, consumer.

Application process
Rolling admission; application on the Lab IX website.

Choosing an Incubator or Accelerator

Many of the hardware accelerators described above are themselves startups. Several are venture-backed, with funding drawn from a combination of strategic (corporate) and traditional venture capital investors. They're constantly refining their models and working to determine the combination of services that best meets the needs of their classes. The authors of this book make every effort to keep the information in this chapter up to date, but you should verify that the program is as we've described before you apply.

New programs launch regularly. Some focus on specific niches within the hardware space. Microsoft, for example, partnered with TechStars to run an accelerator specifically focused on the Kinect (*http://bit.ly/kinect_acclrtr*) and gesture-based interfaces. Another TechStars partnership, this one with R/GA (*http://bit.ly/rga_connected*), helped companies working on connected devices. European-based Springboard ran a program focused on the Internet of Things.

There are regional Asia-focused programs, such as Logistica Asia and HaxAsia. In November 2013, manufacturing giant Foxconn announced plans to start an accelerator in Beijing, though few details have emerged at the time of this writing. University-run programs like the Zahn Center (*http://www.zahncenternyc.com/admission/*), in New York City, offer assistance to undergraduate and graduate students. The broad-spectrum hardware collaboration between TechStars and R/GA appears to be gearing up for a second class.

Offerings are constantly changing, so it's worth searching around to see what's happening in your space a few months before you'd like to join a program.

Given that hardware accelerator programs are relatively new, some hardware entrepreneurs still opt to go the more traditional route and apply to one of the older, historically software-focused accelerators. One of the earliest and best-known of these is Y Combinator (YC), founded by Paul Graham in 2005. Bay Area–based YC provides seed funding, mentorship, and a burgeoning alumni network in exchange for approximately 6 percent of each company's equity. More than 500 companies (http://ycuniverse.com/ycombinator-companies) across 40 markets have come out of the program since its inception. While the program is designed for software startups, the partners have begun to encourage hardware entrepreneurs to apply. In January of 2015, Y Combinator announced a partnership with hardware-focused seed fund Bolt. The Bolt team advises and holds office hours for startups accepted into the Y Combinator program, and YC teams can use Bolt investor Autodesk's Pier 9 Workshop prototyping facility.

Other popular options that have begun to welcome hardware companies are 500 Startups and AngelPad.

Most entrepreneurs coming out of accelerators that aren't specifically hardware focused say that the value of participating in such a program is primarily the alumni network; fellow founders provide a wealth of advice about everything from customer acquisition to hiring engineers.

Pebble Watch founder and Y Combinator alumni Eric Migicovsky says:

> *It's really amazing to join a program and suddenly have links to over 500 founders via the YC email list. Even if there isn't someone who knows the exact answer to your problem, odds are that they probably have a friend or an acquaintance or a past business partner who has worked in that area. We got introduced to our first manufacturer that way; it's an amazing network.*

While a software-focused incubator can't provide access to its own factories, the curriculum and mentors can still offer useful guidance on the business side of growing a company. There is also a credibility boost to coming out of a program that's highly regarded and has an established track record of successful companies.

So, should you give up equity in your company to participate in an accelerator program? Most first-time founders report finding the experience exceptionally helpful. Besides the benefits of participation in the program itself, admission to high-quality accelerators is extremely competitive. The signal of being accepted into one can help with early fundraising or outreach to potential partners or distributors.

That said, it's important to choose a program that truly aligns with the unique needs of your company. Some, such as the TechStars partnership with R/GA, offer unparalled access to digital marketing and design experts...agency expertise that would cost hundreds of thousands of dollars if you were paying for it. If you've already got your CM lined up and are more concerned with building your brand or polishing your UX, that might be ideal for you.

If you've never been to Asia but the economics of your product make it likely that you'll be manufacturing there, a program that will take you to China and open doors to factories may be more in line with what you need. Sproutling, a hardware startup working on a wearable baby monitor, shares its decision-making process in "Sproutling: A Case Study" on page 152.

Sproutling: A Case Study

Sproutling's founder, Chris Bruce, has a long track record of both founding and working at successful software startups. He's got a strong network and knows what it takes to build a thriving company. Even so, when Chris decided to launch his first hardware startup, he opted to go the incubator route. Here, he talks about that decision, and on the experience of being a Lemnos Labs portfolio company.

"When my cofounder Matt and I started Sproutling, I had gotten to the point where I didn't want to work on anything that didn't make a difference," Chris remembers. "I had tinkered with electronics for a long time, and I decided I wanted to build a company that builds smart devices for parents."

In March of 2012, after extensive product-market fit conversations with parents, he built a prototype of a wearable baby monitor. The monitor, worn around the baby's ankle, tracks heart rate, respiration, and movement, as well as room temperature and humidity, and provides insight reports to parents. It can alert parents to health problems but is also useful for what Chris calls "parental time management": predicting when the baby will wake up from a nap.

Chris decided to bootstrap to build the prototype: "This was before hardware had become popular again, and I thought, it's hardware, and it's baby. There's no way I'm getting money." At his previous startup, social gaming company Diversion Inc., he had gone out and raised money before the product hit the market, and he felt that the raise ultimately reduced flexibility and created a massive amount of pressure to succeed.

He also didn't want to turn to Kickstarter too early, because so many crowdfunding raises for consumer products had resulted in teams that hit their goals but still ultimately didn't have enough money to produce the product. New hardware founders, it seemed, consistently mispriced their offerings and delivered late. So the Sproutling founders decided to participate in an accelerator.

"Normally, if I had started another mobile or software company, I would not have gone the incubator route," Chris says. "But I was really worried about the pitfalls of manufacturing and about all of the things that I did not know that I didn't know." He was specifically interested in a hardware-focused program, because he wanted to be able to draw on a network of experts that had shipped consumer devices and overcome manufacturing challenges.

The team applied to Lemnos Labs and HAXLR8R and got into both. The choice between two top-tier accelerators—both with excellent reputations, mentors, and investor networks—was a difficult one. Ultimately, their decision came down to two factors specific to Sproutling's unique situation:

> *First of all, we were really concerned about the industrial design of the product. I didn't know if we'd be able to fully design a quality product and user experience in a constrained three-month period, particularly since the experienced designers were here in the States. We felt like we had more work to do with that before we were ready to get into the manufacturing, so Lemnos's rolling timeline was really appealing. And the second factor is that there's also some stigma about baby products from China. Since this is something that's going to be worn on a baby, we wanted to avoid giving our potential customers any cause for concern.*

The team has been extremely happy with their experience at Lemnos Labs. They appreciate the shared tools and workspace and the community of other founders working across different disciplines. There are accelerator-wide

> Friday barbeques, CEO mentorship days, and workshops that focus on building both devices and businesses. Sproutling has also successfully closed a $2.6 million venture funding round from highly regarded investors. The company is working hard to bring its product to market.

Choosing an accelerator program can be a bit like deciding where to go to college. When you're considering your options, the reputation of the program and the mentor and investor networks certainly matter. But be sure to carefully evaluate your specific needs as a company. Identify what factors are most likely to impact your success, and select the program that is best structured to help you navigate those challenges.

It's difficult to gauge the extent to which a team's success can be attributed to the program it chooses, because startups take years to grow into businesses. However, sources like Crunchbase and Seed-DB (*http://www.seed-db.com/*) rank accelerator quality via metrics such as the amount of funding raised by participant companies and the number and dollar value of successful exits.

If you choose to go the accelerator route, reach out to the program coordinator to learn more about how the accelerator can help your company through the specific challenges you are likely to face; all of the accelerators hold info sessions and are very accessible. Speaking with program alumni is another excellent way to decide if the fit is right for you.

CHAPTER 8

Crowdfunding

When we use the term *crowdfunding* in this chapter, we're talking about *donation-based crowdfunding*: when a group of individuals donate money, often receiving a reward in return, to help a project creator raise the funds needed to bring a product to market (as opposed to *equity crowdfunding*, in which a team fundraises from the public in exchange for equity, but we won't be discussing that here). It's difficult to overstate the impact that donation-based crowdfunding has had on driving the growth of the hardware ecosystem, particularly for teams producing B2C products. In addition to potentially generating enough capital to produce a first run of a product, crowdfunding enables founders to reach an audience of early adopters, grow community, validate a market, get customer feedback and product insights, and generate a substantial amount of buzz.

A crowdfunding campaign is often the public's first exposure to your product and/or company, and it's critical to get it right the first time. Running a successful crowdfunded hardware project will require you to plan and execute strategies for manufacturing, marketing, fulfillment (shipping, warehousing), customer service, returns...it's a ministartup!

The Crowdfunding Ecosystem

Crowdfunding is a relatively new phenomenon. Most donation-based fundraising sites are only a few years old. They themselves are startups that are constantly refining their offerings, adjusting their terms, and developing new ways to serve project creators.

There are dozens of crowdfunding platforms out there, with new ones launching all the time. We're going to start by examining the offerings of the top crowdfunding platforms. Many focus on a specific niche; several skew toward science and research projects, others to art projects, and still others to consumer products. Then we'll move into some best practices that will help you have a successful raise, regardless of which platform you choose.

KICKSTARTER

One of the first donation crowdfunding platforms was Kickstarter (*https:// www.kickstarter.com/*). Started in 2009, its primary purpose was originally to fund art, film, and music projects. Technology and hardware projects began to appear organically. Kickstarter's popularity has grown steadily since its launch; millions of people visit the site every week.

As of April 2015, more than 225,222 projects have been launched on the site, with $1.67 billion raised on the 83,243 successfully funded projects. In other words, there's about a 30 percent chance of success. Kickstarter publishes category statistics (*http://www.kickstarter.com/help/ stats*) on its website and updates them daily.

Kickstarter has some rules (*http://www.kickstarter.com/rules*) that regulate what types (and stages) of hardware products can raise money on its platform. Before launching a hardware product on Kickstarter, it's important to be sure that your project follows these rules. It is possible to raise on Kickstarter in order to produce a prototype; however, if the finished product is one of the backer rewards, then you can't launch with only an idea. You need to include photos and videos of a prototype that conveys the product's current state, along with a production plan and estimated timeline. While CAD drawings and sketches showing the design process are encouraged, photorealistic mockups are not (*http://bit.ly/proto type_rendering*). The team worries that they might mislead a backer into believing the project is further along than it is.

In 2014, Kickstarter made its launch process easier, rolling out a Launch Now feature that allows creators to bypass the previous process of getting approval from a community manager. However, the review by a community manager can be extremely helpful, particularly for a first-time campaign owner. Kickstarter's community managers have often run their own projects in their categories and can be a helpful source of expertise.

Kickstarter requires that a project reach its stated funding goal in order for the funds to be transferred to the project owner. The platform takes 5 percent of funds raised. This doesn't include the additional 3–5 percent in credit card processing fees.

Several notable hardware startups (including Pebble, Oculus, Formlabs, and OUYA) have gotten their start on Kickstarter.

INDIEGOGO

Indiegogo (*https://www.indiegogo.com/*) is another highly regarded platform that a number of hardware startups have used as a launchpad to success. With fewer restrictions than Kickstarter, it's quickly becoming a go-to platform for hardware entrepreneurs. Founded in 2008, Indiegogo has consistently marketed itself (*http://bit.ly/indiegogo_help*) as a platform that allows "anyone, anywhere to raise money for any idea."

To help project owners, Indiegogo maintains an Insights blog (*https://blog.indiegogo.com/*) dedicated to the latest analysis on best practices, including individual case studies and data analysis based on aggregate campaign data. It also maintains the Indiegogo Playbook (*http://go.indiegogo.com/playbook*), which breaks the crowdfunding process down into stages (precampaign, first half of campaign, endgame of campaign, and postcampaign) and provides highly specific tips for each stage. For example:

> *Consider launching on a Monday or Tuesday to help you gain momentum through the week. On average, campaigns launched on a Monday or Tuesday raise 14 percent more in the first week than campaigns launched on all other days of the week.*

Indiegogo's platform offers both fixed and flexible fundraising campaign structures. A *fixed-funding* campaign sets a target amount, and the raise is all-or-nothing. A *flexible* campaign allows the project owner to set a goal, but owners can keep any amount raised. All backers will be charged, regardless of whether the project "tips." That flexibility does come with a higher fee for the project owner; while the Indiegogo cut is 4 percent for a fixed raise *or* for a flexible raise that meets its goal, the fee is 9 percent of whatever amount is raised for flexible campaigns that do *not* meet their goal.

Flexibility is central to Indiegogo's appeal. Campaign owners can set the duration of their campaigns (according to the Playbook, campaigns of about 40 days tend to be the most successful). It's also possible to change perks mid-campaign. This enables campaign owners to tweak and optimize while their raise is live.

Indiegogo recently launched an initiative called Indiegogo Outpost (*http://www.indiegogooutpost.com/*), which lets founders embed an Indiegogo crowdfunding campaign directly into their own sites. This helps founders tailor their raise to their brand and immediately engage backers on their own page, while still benefiting from Indiegogo's campaign management tools. Outpost campaigns also appear on the main Indiegogo site (with Indiegogo's look and feel), so campaign owners can attract backers from among the multitude of users who browse the Indiegogo site.

THE DIY APPROACH

Some founders take the Outpost concept one step further. A small but increasing number of startups opt to roll their own crowdfunding campaign from start to finish. They simply run the process on their own site, independent of the major platforms. Lockitron was one of the first companies to do this. It didn't meet the requirements for Kickstarter's platform, so it built its own crowdfunding framework, Selfstarter (*http://selfstarter.us*), which has its source available on GitHub.

If you're interested in taking this approach but don't feel comfortable writing your own code, there is also an out-of-the-box open source solution called Crowdhoster (*http://crowdhoster.com*), a "WordPress for crowdfunding" currently being developed by Crowdtilt (another crowdfunding platform). Crowdhoster is currently free, and it offers a customized look and feel, integration with Crowdtilt's payments application programming interface (API), and many of the customer management and administration features offered by traditional crowdfunding platforms.

The pros of this approach include lower fees, the ability to have the raise's look and feel fit with your branding, and full flexibility around all terms of the raise. However, it is a lot of work. Scout Alarm, a hardware startup that chose to go this route, shares its story in "Scout Alarm: A Case Study" on page 158.

Scout Alarm: A Case Study

While crowdfunding platforms have undeniably made access to capital less of a problem for hardware entrepreneurs, some founders choose to run a campaign on their own sites. In this case study, the cofounders of Scout Alarm, Daniel Roberts and Dave Shapiro, share why they chose this approach.

Scout is a do-it-yourself wireless home security system. In August 2012, Dan had just purchased a home in Chicago and went looking for an alarm system. He found the market largely archaic and decided to create a more open, affordable, and modern home security system that could connect to personal devices such as computers and smartphones. He and his cofounder Dave began to work out of a Chicago incubator called Sandbox Industries, which gave them a small amount of R&D funding to build a prototype.

Once they'd pinned down a design, they decided to do a crowdfunding campaign to validate the idea. They wanted to use preorder money to help fund the development and offset their initial manufacturing costs. However, they weren't far enough along in development to meet Kickstarter's requirements and felt that they were still too early to appeal to the audiences on most of the other platforms. So they decided to go independent.

Dan and Dave liked the fact that being independent meant they would be able to avoid platform fees and keep a greater percentage of the capital they raised. They also felt they could build their community right on their own site from the start, so they could keep up momentum after the crowdfunding campaign was officially over. Dan says:

> We were able to keep the same site structures up and just convert from crowdfunding to taking preorders and gathering email addresses. Once your Indiegogo or Kickstarter campaign is over, you're done. You have to hope that people who find you after the fact will look for your site and preorder.

Doing an independent raise benefited the team in other ways as well. They worked with the PR person at their incubator and were able to attract publicity through two separate stories: the story of Scout Alarm and an appealing "independent bootstrapper" narrative. They were also able to tailor their raise perks for their customers to a highly specific degree.

While even the most flexible crowdfunding sites require clearly defined perks, Scout was able to ask potential customers about their needs and suggest an appropriate package. Dave says:

> Our product isn't conducive to a one-size-fits-all starter kit. We suggest features based on the number of windows and doors in the home. So we built a UI that would handle that, and we felt that provided a better experience than potential backers would have had

> on Kickstarter. Owning the user experience is a big part of doing an independent campaign.

Going the independent route required a lot of developer manpower. The team used Lockitron's Selfstarter code but still had to build out many of the features they needed. The burden of load testing and site optimization fell entirely on them. They cobbled together suites of existing tools such as Google Analytics and AdRoll, incorporating retargeting to bring people back to their raise. MailChimp and Olark helped them communicate with their backers, Help Scout offered customer service, and Mandrill provided receipt management.

Amazon's Flexible Payments program proved helpful for the independent crowdfunders. "If you're going to use Amazon's Flexible Payments, be aware that Amazon takes a few days to verify corporate bank account information," Dan says. "We were all ready to go and then realized we'd overlooked that step and had to scramble to get it done."

Ultimately, the team encountered many of the challenges of any successful campaign...including fulfillment delays. Using Amazon's system of tokens meant that they had up to a year to charge their customers, and they felt that their community was more patient than they would have been had a crowdfunding platform charged them up front.

Occasionally there were features they wanted and simply didn't have time to build, but the team was happy with the independent experience. "Looking back, there are some things we should have done," Dave says. "I wish we'd had more granular analytics, for example. But overall, we were very happy with how things turned out and would go this route again."

For more of Scout's takeaways on the independent-crowdfunding experience, check out its blog (*http://bit.ly/indy_crowdfunding*).

While there are too many smaller platforms to discuss each one individually, another worth mentioning is Dragon Innovation (*http://www.dragoninnovation.com/*), which focuses solely on hardware. Dragon is a popular project-management partner for hardware endeavors of all sizes. It helps with everything from factory selection and vetting to design reviews to manufacturing support overseas. One of Dragon's offerings is a certification process, which reassures potential backers that the project will be delivered on time. (We'll discuss this certification process a bit more in "Timing with Manufacturing" on page 171.)

Do your research to make sure you're launching on the crowdfunding site best suited to your type of product. Investigate the composition of the audience and the breakdown of the projects. Some founders believe it's important to consider the number of eyes on the crowdfunding site itself. Although you'll ultimately have to drive most of your traffic yourself, launching on a more popular platform might give you a bump if that site has many engaged users or a newsletter with a lot of readers. Kickstarter has blogged about the halo effect of "blockbuster" projects (*http://bit.ly/blockbuster_effects*), such as the Spike Lee (*http://bit.ly/spike_lee_n_kickstrtr*) and *Veronica Mars* raises in the film category. New backers discover the site via these well-publicized products and often back other projects they find while they're there.

Be sure you understand the fundraising structure used by the sites you're considering. Some require you to meet a fundraising goal and charge donors' cards only after the goal is reached. On others, you get to keep any amount you've raised. Check out their acceptable forms of payment, and whether there are any restrictions on foreign backers.

Planning Your Campaign

After you've picked the crowdfunding platform and campaign type that's right for you, it's time to get down to details. The first thing to do is to identify the perks that you want to offer.

UNDERSTANDING BACKERS AND CHOOSING CAMPAIGN PERKS

It's important to understand what makes potential customers excited about giving money to a company for a product that isn't even launched yet. Backers want to feel like early adopters, discovering a cool new product before it even hits the market. They're taking a leap of faith, buying something that doesn't exist yet (and might never exist, in the worst-case scenario).

Danae Ringelmann, cofounder of Indiegogo, believes that people participate in crowdfunding for four main reasons, which she outlines in "Indiegogo Tips and Tricks" on page 161.

Indiegogo Tips and Tricks

We sat down with Danae Ringelmann, cofounder, and Kate Drane, hardware project coordinator, to learn more about Indiegogo's commitment to helping makers and founders run hardware crowdfunding campaigns. Here's what they had to say, in their own words.

When Eric, Slava, and I (Danae) started Indiegogo six years ago, our vision was to empower the world to find what matters to them, whatever that might be. Indiegogo was born out of this frustration around how inefficient, and therefore unfair, finance was. People who got funding typically were the folks who happened to know the gatekeepers and decision makers. If you didn't have access to those gatekeepers, you were pretty much out of luck. The gatekeepers had near-unilateral control over which ideas got born and which didn't. We wanted to build an open platform in which ideas to improve the world could be decided on by the community. There's no application; anyone can participate. Everyone has an equal opportunity to be successful, but that success is based on how hard you work and how much your audience cares.

Based on our experience, people participate in crowdfunding for four main reasons. We call them the *Four Ps*. The first is *people*. People fund people; they're funding the campaign owner or the team behind the campaign. The second is the *project*. Some people just really want the idea to come to life. The third is *participation*. This is a little different in that it speaks more to people wanting to do something interesting, to be a part of the things and the groups and the opportunities that they care about. Maybe someone has always wanted to be a filmmaker, but they're now a lawyer and they have a family and kids; finding a film on Indiegogo and seeing a perk to be on set is their way to participate in something that they couldn't necessarily do on their own. The fourth, obviously, is *perks*: the cool stuff you can get! A platform like Indiegogo is a place where both altruistic and selfish motivations come together in one seamless, cool experience.

For categories that have really started to blossom on Indiegogo, such as hardware and consumer products, we have teams of dedicated support people who work with project owners. Kate Drane and Adam Ellsworth work specifically on hardware, holding office hours and writing up how-to guides. Founders often ask us, "What kind of video should we have?" or "What perk should we offer?" Crowdfunding is still a new experience; there are no established "best practices" yet. As a result, we've worked to make Indiegogo a data-driven company, and we're constantly doing research around what works and what doesn't.

The first key to success is having a great campaign pitch, and first and foremost, that means a video. While they aren't required, campaigns that have them raise 110 percent more than those that don't. We've found that a two- to three-minute video is the sweet spot on Indiegogo.

The second important thing that we tell our project owners is to have a very specific funding target and transparently present how funds will be used. You don't necessarily need to raise all at once; raise what you need to move your project forward. We find that teams with an established track record of execution can come back and raise again, and they'll get repeat funders because they've both built up relationships and demonstrated their commitment to the project. Picking the right funding model is key. On Indiegogo, we have two: fixed and flex. With *fixed* funding, you only get the money if you meet your goal. *Flex* is by far the more popular option and works better for most people. They can keep their project moving forward even if they come up short, and they can put that money to work and come back and raise again later.

Third, make your campaign perks unique. This is where hardware makers can have fun, because crowdfunding opens up the ideation and creation processes to a segment of people who have never been able to be involved before. Consumers typically don't discover devices until they're ready to buy; offer perks that enable them to become part of the creation and ideation process. Create unique items or experiences that you can't get in the store when the device is fully launched.

One thing that's unique about Indiegogo is that we actually encourage perk swapping: perks that aren't being claimed very well can be replaced by a new perk. Our platform is incredibly flexible. A lot of makers use perks to experiment with pricing, test different features, test functions, all because they are able to swap things in and out.

Even if a founder follows all of the rules, sometimes campaigns just fail. That's usually for one of two reasons: either the campaigner wasn't willing to work hard, or the world just doesn't care. On occasion, even if a founder has great perks, proactive outreach, media coverage, etc., the idea just doesn't resonate. This failure can be a really positive thing for a founder. They learn that their idea isn't that good without spending years of time and their life savings developing a product people don't want. They can take that feedback and come back and do a campaign later and be much more successful. One of the most empowering parts of the crowdfunding process is really that market validation.

When you're running a crowdfunding campaign, your goal is to drive traffic to your campaign page, and to convert eyes to dollars. Marketing, particularly by sharing a meaningful story, will drive people to your campaign. The right perks will entice them to support you with funds. Since most hardware founders want feedback from early adopters, your primary perk is probably the product that you're looking to launch.

PRICING YOUR PERKS

A good campaign offers several different *pledge levels*. Research that tracked 75,000 successful Kickstarter campaigns (*http://bit.ly/kickstrtr_campaigns*) showed that $25 is the most popular price for a perk by far, with $50, $100, and $10 fairly common as well. On Indiegogo, the most popular perks are the $25 and $100 tiers. If your primary product is expensive (over $100), you might consider incorporating one or two low-tier rewards that are easy and inexpensive to produce, such as branded stickers or basic t-shirts. These provide an opportunity for people who want to support you to be able to participate. However, be sure that fulfilling these is fast and easy.

One common refrain that we hear from all successful crowdfounding project owners is: "Keep it simple!" Don't offer t-shirts or small tchotchkes unless you have reason to believe that they will genuinely help move your project forward. If they won't, they're a distraction. You'll have to fully understand the unit cost economics for each reward you plan to offer, and you'll have to manage all of those t-shirt orders while simultaneously trying to manufacture the product you actually care about. "We decided not to do low-dollar rewards," says Sonny Vu, founder of Misfit Wearables (learn more about Misfit Wearables's crowdfunding strategy in "Misfit Wearables: A Case Study" on page 165):

> *Most people know why they're visiting your campaign: it's to buy the thing they're hearing about. Unless you're a designer, that's not a schwag t-shirt. If you're going to do a small perk, make it something memorable that enhances the product—say, offering a branded USB cable as part of a life-logging webcam project.*

If you are offering schwag perks, make sure to keep pricing reasonable relative to value. Some percentage of people are happy to kick out extra cash to help a team or idea they feel an emotional connection to, but including perks such as a $25 sticker or $50 mug is a great way to alienate a large chunk of your audience. It will make prices for all of your other perks seem inflated as well.

Misfit Wearables: A Case Study

Misfit Wearables CEO Sonny Vu was previously the founder of AgaMatrix, a mobile health company that makes advanced blood glucose monitoring systems for diabetics. Now, with Misfit Wearables, he's on a mission to "make wearables wearable": to produce fashionable, functional devices with top-notch ambient sensing capabilities. This case study is about Misfit's first product, the Shine fitness tracker, which launched publicly on Indiegogo in November 2012 and is now in stores. Sonny spoke at length about how critical it is to have a well-orchestrated strategy for all aspects of your crowdfunding raise.

Although they chose not to announce it, Misfit Wearables raised a venture financing round in April 2012, months before the Indiegogo campaign. The decision to emerge from stealth with a crowdfunding raise was made because Sonny believes that such platforms can be leveraged for powerful real-time market validation. The Misfit team looked at many options, ultimately choosing Indiegogo because of its flexibility and founder-friendliness.

"This notion that the community size on whatever crowdfunding site you choose matters, it's a myth," Sonny says. "It's a myth because the community sizes are tiny compared to the audience you need to generate yourself. You need to have your own champions."

Champions are the people who will tell others about your crowdfunding raise. Their excitement and strong networks help spur virality and drive traffic and orders. Misfit quickly got to work making a lot of friends. They put together a list of several hundred people, all of whom could be counted on to tell 10 people each. They also orchestrated a concentrated press push:

> The difference between five articles spread across a month as opposed to five articles within a two-day spread is a big difference! Some people say, "I want it spread out so that you keep hearing about us." No. You will just not have heard about us, you will not hear about us again a few days later, and so on, however many times that happens.

If press is clustered, it will perpetuate. Five articles across two days is much more likely to engender 20 more blog posts, and that attention will trickle down to Facebook, Twitter, and other social networks, where it will spread further. The Misfit team felt that they couldn't afford a good PR firm, so they reached out to the press through warm introductions from friends and investors, and began to develop relationships. Most reporters and bloggers they reached out to were willing to embargo a story until the day of the launch. The team expanded beyond the traditional tech and gadget blogs, and reached out to mainstream publications that appealed to their target market segment: customers who purchased premium brands.

Once the press and the champions were established, Misfit got to work on timing. They wanted both of these sets of supporters to feel special, so they opened the crowdfunding campaign at a time when the champions could get in first. Then they lifted the embargo, and the stories went out. Only after these things were done did Misfit send out a broader global press release. Sonny adds, "It cost us $3,000, or some outrageous amount, but the agency we used did translation into 10 languages. We found that 32 percent of our coverage came from overseas. Don't neglect the international market!"

The same degree of precise orchestration went into the design of the Indiegogo campaign itself. Although they decided not to spend money on PR, they chose to hire a team to produce a highly polished video, which generated 650,000 views. Misfit carefully studied completed campaign pages (both successful and not) and observed that looking at Indiegogo's site on a regular-sized monitor showed approximately two perks. They wanted to get the most important reward, the Shine, above the fold.

Because Indiegogo offers project owners flexibility with rewards (it's possible to increase or decrease prices, quantities, and offerings throughout a campaign), Misfit was able to test new offerings as the campaign continued. They experimented with prices, charging higher prices for special colors. They read user comments about the perks they were offering and added new perks (including a popular necklace add-on) in response.

Misfit's successful crowdfunding raise was the culmination of several months of organizing and strategizing (see Figure 8-1 for its Indiegogo campaign timeline and Figure 8-2 for an expanded view of the launch-day schedule). However, although a great deal of time and effort was devoted to producing a top-notch campaign, the team never lost site of the ultimate goal: to introduce, and sell, the Shine.

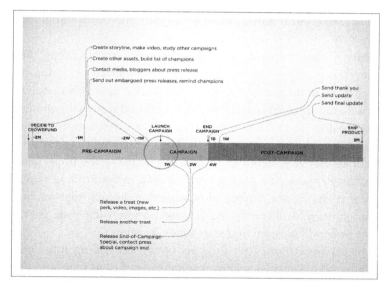

FIGURE 8-1. Misfit's Indiegogo campaign timeline (courtesy of Sonny Vu)

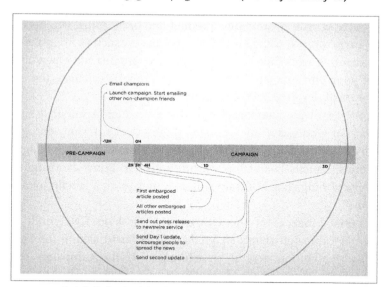

FIGURE 8-2. A detailed schedule of the launch day circled in Figure 8-1

Pricing your perks appropriately affects not only your crowdfunding campaign success; it is important to the success of your overall business as well. Chapter 10 discusses product pricing in more detail. If you're opting to do a crowdfunding campaign as your public unveiling, read that section.

Many hardware founders who launch on a crowdfunding platform significantly underprice their products. They don't factor in the margin required to be profitable on the other distribution channels that they'll be selling on after the campaign. The public is introduced to the product at a low price point...and then sees it costing 25–50 percent more in stores or on the company website a few months later. This might make early backers feel like they've gotten a great deal, but it creates the impression that you've raised prices. It's great to reward your crowdfunding supporters for taking a chance and buying something that doesn't exist yet, but pursue this strategy deliberately. Be sure you know what your postcrowdfunding price will be *before* you start your campaign.

Inexpensive schwag isn't the only option for adding more perks to your campaign. Consider adding special treats to the high end as well. High-tier rewards that incorporate backers directly into your process can be appealing to crowdfunders who want to be involved in the development. A backer can help name a product line or be added to a beta-tester list for early runs. You can also offer experiences, such as an in-person or Skype demo, a workshop on how to use the product, or a personalized tour of your office or lab. Just be sure to properly calculate the cost of these types of rewards...including your time!

Some hardware companies offer their product as either a kit (low-price tier) or a fully assembled version (high-price tier). Consider offering limited-edition versions or colors of your core product, or campaign-only accessories; scarcity is a very powerful motivator. Some campaigns offer *stretch goals*: things that the team will do if they reach a certain total dollar amount.

Otherfab, makers of the Othermill (a desktop CNC mill) ran a Kickstarter campaign (*http://bit.ly/othermill_kickstrtr*) asking for $50,000 but also mentioning several stretch goals. If the campaign reached $100,000, the company would create a job for a machinist. At $250,000 it would expand its software offering ahead of schedule. At $500,000 it would be able to create a second production line, so it could ship all of the mills much more quickly. The project raised $311,657.

While creativity might tempt you to offer a ton of different rewards, don't overdo it. The average for a successful project on Kickstarter is nine. If you're on a platform with a flexible perk structure and something isn't selling well, you can adjust perk pricing or offerings while the raise is ongoing. This enables the founder to experiment with different feature sets and price points, and to use the crowdfunding process for deep market exploration.

CREATING A FINANCIAL MODEL
Pricing your reward tiers is a challenging exercise in being cognizant of how much money supporters are willing to spend while still pricing in an appropriate amount of *gross margin* (the selling price of an item minus the cost of goods sold)—particularly if fulfilling this campaign is going to be your full-time job for a while. (It's important to note that this section touches on the pricing considerations necessary for successfully executing a crowdfunding campaign. Chapter 10 includes additional pricing exercises relevant to growing a company. If you are planning to distribute your product through big-box retail stores at some point in the future, for example, you must build in enough margin *now* to remain profitable while keeping your price consistent across all channels.)

For now, we're going to assume that you know what the total wholesale cost to manufacture each individual widget is and work through the additional economic considerations specific to determining *cost of goods sold* (COGS = cost of products, including production cost, freight/shipping, labor costs, factory overhead expenses, storage) per item for a crowdfunding platform raise. If your product is still in the concept phase, you must do extensive research on production before you initiate a crowdfunding raise (and be sure you've read Chapter 5 and Chapter 6).

Presumably, you have a fundraising goal in mind. You'll definitely need one for a fixed campaign. You want the total amount raised across all reward tiers to move your project forward. Crack open Excel or other spreadsheet program of your choice, and model out your *burn rate* (the amount that your startup spends each month). Figure 8-3 shows a basic example of a crowdfunding perk pricing worksheet (download from this book's GitHub repository (*http://bit.ly/crowdfunding_perk*) to use it yourself).

	A	B	C	D	E	F	G
1			Number of employees	1	3	5	7
2			Economy-of-scale discount	0%	2%	5%	8%
3	Assumptions per unit:		Number of Units	100	1000	5000	10000
4			Production costs:				
5	product cost per unit:	$ 30.00	Product	$ 3,000	$ 29,400	$ 142,500	$ 276,000
6	packaging/inserts per unit:	$ 3.00	Packaging	$ 300	$ 3,000	$ 15,000	$ 30,000
7	shipment costs from mfg to you, per unit:	$ 5.00	Shipment (from factory)	$ 500	$ 5,000	$ 25,000	$ 50,000
8	certification fee (flat):	$ 5,000.00	Certification Fee	$ 5,000	$ 5,000	$ 5,000	$ 5,000
9	customs entry bond:	$ 300.00	Import Fees	$ 300	$ 300	$ 300	$ 300
10	HTS tariff rate:	3.50%	Tariffs	$ 116	$ 1,134	$ 5,513	$ 10,710
11	*please see notes for clarification		MPF & HMF Fees	$ 898	$ 4,535	$ 20,173	$ 38,735
12	cargo insurance rate:	0.30%	Cargo Insurance	$ 10	$ 97	$ 473	$ 918
13	other fees (all-in amount):	$ 5,000.00	Misc Expenses	$ 5,000	$ 5,000	$ 5,000	$ 5,000
14			Fulfillment costs:				
15	shipment cost to customer:	4.95	Shipping to customer	$ 495	$ 4,950	$ 24,750	$ 49,500
16	handling fee per unit:	0.50	Handling	$ 50	$ 500	$ 2,500	$ 5,000
17	monthly storage per unit:	0.50	Storage	$ 50	$ 500	$ 2,500	$ 5,000
18	defective rate:	10%	Defectives (assumes refund + return shipping)	$ 1,050	$ 10,495	$ 52,475	$ 104,950
19	hourly wage:	12.00	Customer Support	$ 1,200	$ 3,600	$ 6,000	$ 8,400
20	hours needed to fulfill (per employee):	100					
21			Crowdfunding economics:				
22	perk cost to customer (incl s&h):	$ 100.00	Cash from Perk Sales	$ 10,000	$ 100,000	$ 500,000	$ 1,000,000
23	crowdfunding fee:	5%	Crowdfunding Platform fees	$ 500	$ 5,000	$ 25,000	$ 50,000
24	card processing fee:	5%	Payment Processing fees	$ 500	$ 5,000	$ 25,000	$ 50,000
25			Gross Margin	$ (8,967)	$ 16,489	$ 142,818	$ 310,487
26							

FIGURE 8-3. A crowdfunding pricing model worksheet

Besides the cost to manufacture the goods and the cost of your team's labor, you should be taking into account the following potential expenses *for each perk*:

- Product packaging
- Certification fees
- Warehousing/storage costs
- Packing and shipping (including labor costs)
- Defectives (including return shipping and replacement costs)
- Customer support staff
- Import/export costs (tariffs, duties, etc)
- Fees (to the crowdfunding platform, possibly also to the payment processor)

While the majority of bullets on that list are standard for any hardware business, *fees* require special consideration when you do a crowdfunding raise. Crowdfunding platforms charge fees. Most take a 3–10 percent cut of donations received. As mentioned earlier, Kickstarter's take is 5 percent. Indiegogo's is 4 percent for fixed-funding campaigns and 9 percent for flexible funding. Most platforms have an additional 3–5 percent charge for processing and transfer fees on the credit card payments they're collecting for you.

Some of the variability in the credit card processing fees comes from the location of your buyers, which can be difficult to predict. So the minimum price for each widget must include the cost to manufacture the widget itself plus enough to cover the fees levied by the platform you choose. Be aware that the bank might charge additional fees to receive disbursement, depending on the type of account you have.

It's difficult to know what kind of demand you'll see once you're live, so modeling for different scenarios is important. Pinoccio, an open source microcontroller for building the Internet of Things, raised $105,200 on Indiegogo. The team posted an excellent writeup (*http://bit.ly/pinoccio_projection*) featuring an Excel template that models out potential crowdfunding profits and costs given backer assumptions ranging from "Meh" to "Unreal." You can download the Excel template from Pinoccio (*http://bit.ly/pinoccio_sprdsheet*) to use as a jumping-off point for your own financial modeling.

It's important to decide if you're going to offer a limited quantity of reward units at each tier, or if you'll let the market dictate how many you'll make. If you don't want to put a limit on your number of rewards, model out several scenarios for manufacturing and fulfillment. While your bill of materials might drop as you scale up the number of widgets, be sure to appropriately price in the additional warehousing and manpower costs to fulfill the orders. If you're determined to do a small exploratory run and handle fulfillment yourself, placing a limit makes sense: A wild, runaway success of a campaign might leave you having to produce more units than you can plausibly make and store in your garage.

TIMING WITH MANUFACTURING

Set a timeline for your crowdfunding raise that appropriately syncs with your manufacturing timeline. There is often a 14-day wait between the time that a crowdfunding campaign is completed and the date that the platform distributes your money; make sure you end the campaign with enough time for the money to hit your bank account before you have to pay anyone.

Before any campaign, you should have your manufacturing partner identified and lined up. You want to give your backers a reasonably accurate date on which they can expect to receive their perks, and you can't do this if you have no idea who's going to make your product. Sonny Vu of Misfit Wearables says:

> *A lot of the reason why folks are late to deliver on crowdfunding projects is because they're first-time makers. And ultimately, to make a product versus making a product that can be made are completely different skill sets. From the very beginning, you must figure out if what you're offering can be made*

in scale. Our first runs were always something like 100 at a time, and they had to be done within a day or two. If it can't get done in that timeframe, it cannot be made. Make sure to figure out the feasibility around manufacturing and scale for all of your perks from the beginning.

A crowdfunding raise requires months to prepare for. You can come up with the most exciting perks and have the most well-orchestrated marketing push in the world, but what will ultimately matter most for your company postcampaign is your ability to deliver. When potential customers search for you, you don't want the narrative that comes up to be one of repeated delays and dissatisfied backers. Have long conversations with your supply chain before starting your campaign, and be sure you have a very solid sense of their capabilities.

Many of the crowdfunding sites let you add on perks as your campaign progresses. One option for handling runaway demand is to have perks that specify different delivery dates. If your factory can produce only 5,000 units for you in January, create a perk that promises January delivery and set the availability at 5,000. Then, if that perk sells out, open up a new reward with a later delivery date.

Dragon Innovation, a manufacturing services company that runs a hardware crowdfunding platform of its own, recently launched a product that can help entrepreneurs vet their manufacturing and supply chain processes. Called Dragon Certified, the program includes detailed DFM and assembly reviews, COGS analysis, and the creation of a manufacturing strategy. The Dragon team also helps with the crowdfunding materials themselves, reviewing incentives (including ensuring that perks are priced appropriately), networks, and campaign materials. The fee for this service is $5,000, plus 2 percent of the funds raised from a successful campaign (the campaign can take place on an external crowdfunding platform). Entrepreneurs who have their products certified often continue working with Dragon to scale manufacturing as they grow their businesses postcrowdfunding.

Campaign Page Marketing Materials

Now that you've chosen rewards and worked through the pricing process, it's time to start crafting your pitch. A successful donation crowdfunding raise involves producing top-notch marketing materials (copy, video, photographs) for your page, and planning an outreach or PR strategy to drive traffic to the campaign. You have to be able to tell a great story and make sure people are hearing it.

The mantra to keep in mind here is "show, don't tell." A picture is worth a thousand words, and a video is worth even more than that. While most backers will read your profile copy, their eyes will be drawn first to the images and video that you provide. High-quality photos of your prototype are a must. If you don't have a design prototype, professional-quality sketches or CAD drawings can help people visualize what they're donating toward.

Kickstarter's FAQ points out that projects with a video succeed at a much higher rate (*http://bit.ly/kickstrtr_faq*) than those without (50 percent versus 30 percent) and tend to raise more money. On Indiegogo, 64 percent of campaigns incorporate video. However, creating a good video can be tricky. If it's too short, it won't convey your story; too long, and viewers will simply stop watching. The average video length for an Indiegogo campaign video was 3 minutes and 27 seconds in 2012, but Indiegog's data scientists found that the "sweet spot" was 2.5 to 3 minutes (see "Indiegogo Tips and Tricks" on page 161 for tips on the most important elements of an Indiegogo campaign pitch).

John Dimatos, Kickstarter's Project Specialist for Product Design and Technology, confirms that two to three minutes was optimal. "It's the right amount of time to grab people's attention and tell them what the product is," he says.

An ideal video is clear and concise, sets a tone for the campaign, and makes a personal appeal. While showing off your product is important, don't spend the whole video pitching each feature. Instead, tell your story. "Don't bury the lede," John advises. "A lot of people try to describe who they are and where they came from and why they're doing this, and *then* they get to the product. You can just show what you've made right up front. Get it out there. Then go back and fill in the details."

Since people are donating to support your team, do be sure to feature the team after you've highlighted the product. Emphasize the collaborative nature of crowdfunding, and ask them to join you on your adventure (include a call to action!). If you have a particularly exciting perk for the crowdfunding raise, mention it. Close with your product or company URL, so that those who want more information have somewhere to go.

In addition to beautiful images and video that will make people want to own your widget, you will need well-written copy that tells who you are and why your donors should want to help you. This is the story that will pull it all together. It should communicate your passion for the project, what you hope to achieve, and why you believe this widget is going to make backers' lives better. Be honest and transparent. Talk about how you plan to use the funds. Describe your product and key features. If you're launching a product in a crowded space, consider including why your product is different from what's already available.

Be aware of the power of white space; don't confront your readers with a long wall of text. A good campaign writeup strikes a balance between text and pictures. Describe what you're making right at the top. "Know who your audience is," John says. "If it's early-adopter technology types, they typically need less hand-holding and will break away from the written description early to check out the perks." So get the important information up at the top. Other consumer groups might be more traditional; they will continue on and read about all of the features. Much of the content you create for your profile can go on your company's landing page, too.

There are many resources out there for creating the best possible campaign marketing materials. If you're not a natural writer, consider drafting your copy and then hiring a professional editor or copywriter through a freelancing site such as oDesk or Elance. This doesn't have to be expensive; if you shop around, you can find editing at reasonable rates. Editorial services like The Word Spa (*http://wordspa.com/*) can help polish your prose and advise you on writing crisp, sparkling copy.

To find a professional for photography or video production, identify the campaign videos that resonate with you; many include production credits. Some video marketing experts, such as Sandwich Video (*http://sandwichvideo.com/*) (clients include Coin, Jawbone, Square, and many more top tech companies), work with startups and have produced many crowdfunding campaign videos. Sandwich's How it Works page (*http://sandwichvideo.com/how-it-works/*) describes its creative process and turnaround time; it takes six to eight weeks to produce a high-quality video, and top-quality work isn't free.

If your budget simply doesn't permit hired help, don't worry. John says, "Regular DIY videos do just as well as professional. There's no *one true path*." If your product is best represented by a slick, well-produced video—say, if the product is a luxury item—then work to achieve something polished. If you have a more DIY feel or are reaching out to an audience of makers, a low-production video might resonate more. According to John, "Videos should showcase the product personality."

There are many free resources that offer guidance on how to make a good video without professional help. Indiegogo has an excellent blog post (*http://bit.ly/pitch_vid_tips*) with helpful tips, and Kickstarter has written on the subject (*http://bit.ly/kickstrtr_awesome_vid*) as well. Some media and advertising companies that work specifically with very early startups will reduce their fee if they love the product, or work out an equity or revenue share deal instead.

Driving Traffic

So now you have chosen a platform, set a timeline, and created marketing materials. It's time to develop a strategy for getting people to your campaign page. Regardless of what crowdfunding site you choose, you will be responsible for driving most of your traffic. For many founders, getting the word out is the most difficult part of the crowdfunding process.

LEVERAGING SOCIAL MEDIA AND EMAIL LISTS

Leveraging social media is key to a successful crowdfunding raise: it's free, and most crowdfunding platforms are well integrated with social channels, so you can easily promote your project and push updates. If you don't already have a social media presence, begin to build one several weeks or months ahead of your launch. Identify the social sites that matter, and create company accounts (and, if you're comfortable, personal accounts). Twitter, Facebook, and Google+ are the Big Three, but Pinterest, YouTube, and Instagram can be powerful tools for distributing images and video clips.

Start by sharing your company's social presence with your friends and family. Ask them to share your campaign with five friends of their own. While you might feel somewhat uncomfortable to market to (and through) them, these are the people who are mostly likely to believe in your ability to succeed. Before you push your campaign out to the world, ask your personal network for feedback. Danae Ringelmann of Indiegogo says:

> What we see is that the most successful campaigns typically raise 20–30 percent of their target from their inner circle first. Using platform social media integration, we enable them to reach secondary and tertiary circles.

And then, if the founder is doing a really good job of keeping their audience engaged, we'll start to promote the campaign ourselves on our channels. We think that campaigns should get early validation from people that they know: friends, other customers they've had, people who know they have a track record of success. Build from friends to friends-of-friends, and then move on to strangers.

Many sucessful crowdfunding campaigns begin building up an email list months ahead of their launch. You'll need a website or landing page prior to launching the campaign (because you have to drive traffic to your own site to keep up momentum after the raise is over...more on this later!). The sooner you get a site up, the sooner you'll be able to begin collecting email addresses from interested potential customers. After you have the list, you can use a service like TowerData to extract demographic data and learn more about your audience.

Pebble, the connected watch that raised $10 million on Kickstarter, began its campaign by reaching out to the 6,000 people on its email list (see more about Pebble's crowdfunding strategy in "Pebble Watch: A Case Study" on page 176). Pebble had been building up the list for years. Although its runaway success was attributable to many factors, it's worth noting that only 25 percent of its backers were Kickstarter members before the Pebble. The remainder came from traffic driven first by the team, and then organically by the press buzz surrounding the project. If you have sponsors or partners, ask them if they'll promote your raise on their lists as well.

Pebble Watch: A Case Study

Pebble founder Eric Migicovsky ran a Kickstarter campaign that raised more than $10 million, with the first million in just over a single day! When he talks about how he did it, he emphasizes the importance of building a network over time, getting your product out there (however rough it might be), and getting feedback as early as possible.

The idea for the Pebble watch (then known as inPulse) came about in 2008, when Eric—an avid cyclist—grew tired of being disconnected from important messages while riding his bike. He hacked together a small gadget using an Arduino, a display from an old cell phone, and some parts that he bought on SparkFun. The original design wasn't a wearable; his plan was to make a bike computer.

The first inPulse had a Bluetooth chip that could talk to a smartphone and receive emails and text messages. Eric paired it with his Blackberry. Friends got excited about his project and started encouraging him to turn it into a company, and the Pebble was born. A few iterations in, he changed the form factor to a wearable watch. While the team developed the hardware, they put up an information page and started collecting email addresses from interested potential customers.

Two years after the first prototype, the inPulse watch hit the market. Sales were handled through the company's own site. Eric and his team were upfront about the beta state of the product. They emailed people who'd signed up to indicate interest and advised them that they could purchase a beta version. Despite the early stage, many people still wanted to buy it. The team borrowed money from Eric's parents, produced a small run of just over 100 watches, and shipped them out to their first customers.

"I always like to tell people to start small," Eric says. "Worry about how to make 10, 50, 100 of your thing. Get out there, get feedback...and then worry about making more." Producing smaller runs gave Pebble insight into where problems were likely to arise. Shipping a product to excited early adopters helped them learn more about what their users wanted.

Four months after starting the beta, Eric applied to Y Combinator (YC), a top Silicon Valley incubator. At the time, he knew nothing about running a startup, and YC rarely accepted hardware companies. To his surprise, Pebble got in. The team found the experience incredibly valuable. Suddenly they were part of a network of 500 founders who could help with advice on all aspects of building a company. Unfortunately, they found themselves unable to raise venture funding following the completion of the program. So they turned to Kickstarter.

The Pebble team spent six weeks planning their Kickstarter campaign. They improved the company's website, putting up information to clarify how the new version of the watch was an improvement over the beta. They identified the features that their audience found most exciting—the notification system, the sports and fitness capabilities, and the fact that the user could customize the watch face—and came up with a concept for a Kickstarter video that emphasized these features. They made the video themselves, because they felt that they knew the product better than anyone. They decided to keep things simple and limit their reward offerings to just the watches. Finally, on April 11, 2012, it was time to go live.

Recognizing the need to drive traffic to their campaign, they first turned to the customer email list they'd built up over the previous five years. They emailed all 6,000 people on the first day of the campaign, letting them know that they finally had the opportunity to support Pebble and prepurchase a watch. They also granted a press exclusive to the senior mobile editor of Engadget, who'd written about the company before. She wrote an article about the raise that ran on the morning of the launch.

Eric also agreed to speak with other reporters about various facets of their campaign. Kickstarter itself was becoming wildly popular at the time, so a number of those interviews focused on the concept of crowdfunding in general. Unexpectedly, the Pebble campaign went viral, taking in close to $1 million on the first day. Almost 75 percent of the supporters were first-time Kickstarter users. Within three days, Eric had to hire a PR agency to help manage the attention.

As the Kickstarter campaign took off, the Pebble team decided to shift manufacturing to China. The team had originally been hoping to sell 1,000 units; they sold 85,000. The runaway success did present some challenges in terms of fulfillment, but the team had several years of experience with manufacturing by this point.

"While every case is different depending on manufacturing techniques, a good general rule of thumb is once you need to make somewhere around 3,000 to 5,000 units, it's more efficient to go overseas," Eric says. "Manufacturing is hard. I think it's probably best to go on Kickstarter and sell a hardware product after you've already made and shipped an existing physical hardware product."

Again, the Pebble team relied on their network and reached out to contract manufacturers who came recommended by friends and colleagues. One of the most helpful resources was Dragon Innovation, which had extensive experience in the Chinese manufacturing ecosystem.

Pebble broke Kickstarter's previous record of $3.3 million raised with a month remaining and ultimately hit $10,267,845 from 68,929 backers. They were able to raise venture funding, closing a $10.3 million round in December 2012. The company continues to grow. Currently, the watch is available at Pebble's site, Amazon.com, Best Buy, and others.

CONNECTING WITH THE MEDIA

Moving on to strangers involves reaching out to publishing channels and online communities. You researched your target market prior to product development; now is the time to reach out to them where they are. Mommy blogs? Cooking sites? Car aficionado forums? Start spreading the word about your crowdfunding campaign on the message boards, forums, or online communities where your niche customer gathers. If you're producing a premium lifestyle product, reach out to fashion and lifestyle blogs and magazines. While an increasing amount of media is consumed on blogs and online-only publications, a feature or mention in a newspaper or magazine is valuable if the audience is a fit.

Reach out to influencers or community leaders. If you're able to post directly to a community yourself, be aware of community culture and rules about promoting products. Reddit is one example of an online community with highly engaged users but little patience for excessive self-promotion. Indiegogo provides some excellent tips for sharing content and engaging on Reddit (*http://bit.ly/engaging_reddit*), including using promoted posts.

Coverage from industry publications or community blogs drives traffic. Many journalists and bloggers who write for these sites are receptive to genuine, authentic pitches from someone whose story will be interesting to their audience. Find the ones who have covered your general space in the past, and reach out.

Celery (*https://www.trycelery.com*), a startup that helps hardware startups accept preorders, has an excellent free PR guide (*https://legacy.trycelery.com/shop/pr-ebook*) for startups and crowdfunding campaigns. Written by former *New York Times* reporter Chris Nicholson, it discusses the "Three I's" that motivate journalists and can help you establish a productive relationship: information, introductions, and ideas.

Journalists are always looking for useful information about topics they care about. Some ways to give them that information include providing new data, answering a question, or alerting them to a new technology. You could offer introductions by opening up your network and putting journalists in touch with potential sources for a story. You might even offer a pitch for a story, a way of thinking about a connection between industries, or a heads-up about a new trend in your market.

In PR outreach, like anywhere else in business, relationships help. Reach out to writers with a warm introduction by someone you know wherever possible. It might take time to get to know the people who can help you. However, if you plan to turn this product into a company, you'll have to build relationships with the press and online marketing channels in the future. Consider the crowdfunding campaign your public unveiling, and begin to build those relationships before it happens.

Be aware that "We're doing a crowdfunding campaign!" is no longer a compelling story. Crowdfunding itself is no longer a new concept; hundreds of thousands of campaigns have happened since Kickstarter and Indiegogo launched. You have to pitch more of a story than that. The story should be about your product, your journey, your solution to a huge problem...something that will generate an emotional response from a reader. It should not be about the fact that you're raising money.

Chris Nicholson stresses the need for emotional resonance:

> *Every story that matters has some emotional consequence, and emotions will hook readers. No emotions, no readers. Find the parts of your story that elicit feeling, or that would make someone pause with surprise, and include those in your pitch. That's how you convince reporters that your story is newsworthy. It has to be something that moves people so much they want to share it.*

The more positive coverage you can get, the better. Misfit Wearables, makers of the Shine activity tracker, carefully lined up a team of top-tier reporters willing to release articles all on the same day. They felt that a "blitzkrieg" approach, with a volley of posts hitting at the same time, would increase their chances of getting noticed, thereby perpetuating more coverage from other blogs, and more organic social shares (see "Misfit Wearables: A Case Study" on page 165 for more details).

ORGANIZING PR MATERIALS

Regardless of the specific media outlets you choose, you'll need the following materials before kicking off your raise:

Pitch email
> Choose an email subject line that attracts the reporter's attention: if you are cold-emailing someone you don't know, be sure to (briefly!) articulate why they should want to write about you. Show that you're aware of their beat and chose them for a reason. Keep in mind that the story you're pitching *isn't* your crowdfunding campaign. It's the launch of your product. The mechanism you're launching with is a small detail. A reporter's job isn't to help you raise funds; it's to tell

her audience about a product that interests them. So your email to the reporter (or blogger) should absolutely not include a plea for help with fundraising.

Media timeline

A media timeline helps organize when you will reach out to the press, and when you would ideally like various articles to appear. Plan well in advance. It can take a few weeks for a reporter to have the time to write a story about you. If you want to *embargo* the announcement (request that it not be published until a certain date), clear that with the reporter in advance. Social media outreach and content should be scheduled on your media timeline as well. Hootsuite, Buffer, and similar tools are very useful for scheduling specific tweets and posts.

Press release

Press releases are a bit formal, but they are a way to reach many journalists at once. The press release should describe what you're building and why it's exciting, and it should include a few quotes that reporters can use if there isn't time for a personal interview. (Note that simply emailing around a press release is not likely to be successful if the reporters you send it to have never heard of you.)

Don't neglect the international market! Translate your press release into other languages before it hits the newswire. Almost 45 percent of Pebble's orders came from outside of the US. For Misfit Wearables, it was 32 percent. Unless the shipping process is prohibitively difficult or costly, or your product is subject to specific local certification requirements, you can use the crowdfunding campaign as an opportunity to build some buzz overseas as well.

Be mindful of your broader audience as you craft your pitch email and press release. Chris Nicholson of Celery advises:

Never use marketing lingo or business jargon with the press...at the end of each sentence, read what you've written and ask yourself: Would a normal human being say that in conversation with a stranger? If the answer is no, go back and rewrite it.

Many first-timers wonder if they should hire professional PR. There's no right answer to that question. Top professional public-relations representatives are experts at creating emotionally resonant stories and garnering media coverage for their clients. Deciding whether or not to hire a professional is a function of budget and how well-connected the team already is. Fees vary widely across the industry. Some retainers are in the $2,000 range, while others go into the low tens of thousands of dollars.

PR doesn't work miracles. Fundamentally, your product still has to be high-quality and in demand. Extensive conversations with potential customers during the development phase is the best way to ensure that your project will resonate. According to Kickstarter's John Dimatos:

> *You don't have to hire PR to be successful. It depends on who you are and what makes sense for you. There is no one true path that every tech project needs to take. That being said, the most important thing that an entrepreneur can do to attract attention is to make sure that they have a good product. It may sound simplistic, but putting energy behind sincerely communicating what makes your product great will always win over tricks and gimmicks.*

Second-time founder Sonny Vu of Misfit Wearables skipped professional PR and opted to make use of his own network of connections, reaching out to reporters and asking friends to do the same. If you are a first-timer, don't have many connections, and feel more comfortable hiring a professional, you can expect a PR firm to cost in the neighborhood of $10,000 to $20,000 a month. This typically includes developing messaging and generating launch coverage.

Wareness (*http://www.wareness.io/*)—which helped launch Coin, Ringly, and Tile—is one example of a PR firm you might consult to represent you. Focusing on hardware startups, Wareness goes a step beyond traditional PR and acts as a full go-to-market consultant. This process includes involvement in crowdfunding campaign design, positioning, a launch strategy, and media/analyst relations.

Wareness accepts approximately 2 percent of inbound clients. It begins its relationship with a founding team approximately three months ahead of the campaign launch target date. This is the amount of time that Wareness feels is necessary to develop the campaign and do media outreach and strategic positioning. Enzo Njoo of Wareness advises:

> *Ahead of reaching out to a PR agency, teams should raise enough funding to launch their campaign and establish partnerships with video producers*

and web designers, since these marketing assets tend to take the most time to develop.

Wareness encourages companies considering crowdfunding to raise enough money to support their campaign before reaching out for PR. Ideally, this would be a raise of approximately $1.5 million, including a marketing budget of approximately $300,000.

> **TIP** This process (raise funding first, crowdfund second) is an approach most commonly taken by teams that intend to use crowdfunding as their go-to-market strategy. Chapter 9 discusses this approach.

Besides media coverage, another way to drive traffic to a crowdfunding campaign is to pay for it. Advertise. Buy Google AdWords, look into Promoted Tweets, or pay to be featured in the Facebook News Feed. Advertising isn't free, so if you have a limited budget, this might not be feasible for you. (See Chapter 10 for a discussion of digital marketing strategies.)

While Your Campaign Is Live

Congratulations! You worked through your comprehensive plan to launch a crowdfunding raise, and now you've hit the "Go" button! It's time to move into the management phase: engaging with the community, tweaking the campaign, and keeping up momentum.

Indiegogo's blog advocates aiming to raise a third of your fundraising goal within the first quarter of your campaign, so successfully executing on a media strategy in the early days is very important. But even after that initial buzz dies down, there are strategies that can help you ensure that dollars continue to roll in.

DATA-DRIVEN CROWDFUNDING AND REAL-TIME ADAPTATION

Several of the big crowdfunding platforms include *analytics*, which will tell you a lot about where your audience is coming from. Indiegogo and Kickstarter both offer dashboards (see Figure 8-4 for Kickstarter's dashboard) that let project owners identify the blogs, social media platforms, and individuals that are their best source of referrals.

FIGURE 8-4. The dashboard overview

The Kickstarter dashboard, for example, shows what percentage of your traffic is coming from inside (a user browsing multiple projects on Kickstarter) versus outside of Kickstarter, as shown in Figure 8-5. If the user came to Kickstarter specifically for your project, it tells you the site that sent them (Engadget, Mashable, etc.). You can see how many backers came from each site, as well as the total amount of money they pledged to the project and which rewards are most popular (as shown in Figure 8-6).

Referrer	Type	# of Pledges	% of Dollars	Dollars Pledged
Direct traffic (no referrer information)	External	216	20.85%	$30,826.50
Twitter	External	86	7.45%	$11,007
Engadget.com	External	83	8.09%	$11,952
Search	Kickstarter	79	7.49%	$11,086
Techcrunch.com	External	75	6.65%	$9,830
Technology Category Page	Kickstarter	70	6.43%	$9,504
Embedded Project Video	Kickstarter	52	4.74%	$7,000.11
Supermechanical.com	External	46	4.13%	$6,111.01
Facebook	External	46	3.74%	$5,521
Google.com	External	40	3.65%	$5,684.99
News.ycombinator.com	External	32	3.42%	$5,055
Theverge.com	External	29	3.09%	$4,563

FIGURE 8-5. Traffic sources

John Dimatos from Kickstarter describes how to leverage the information in the dashboard to generate more traffic:

> You'll see traffic coming in from sites that you didn't contact. Contact them! Talk to them. Find out who they are and how they found out about you, why they recommended you to their audience. Ask them who else they think you should talk to. Use the attention it as a way to build your

network. It's also worth researching what other websites are like that site, and reaching out to them as well.

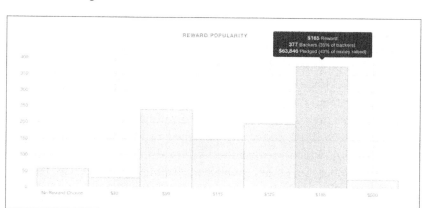

FIGURE 8-6. Reward popularity

You can use analytics to get a sense of how well your video is performing, how long people watch before dropping off, and whether people are watching it on your campaign page or in an embedded viewer on another site. If you share your video on YouTube, examine the analytics there. Set up Google Analytics for your company page. Check campaign views relative to campaign contributions to get a sense of your conversion rate and determine if perhaps you should be adjusting your marketing materials or increasing outreach.

When you send out emails, send them using a program like MailChimp that tracks open rates. You can write emails in advance and schedule them to go out at particular times of day or days of the week. Weekends are not optimal; Indiegogo has found that most participants fund campaigns during the workday.

When you include links in any social media outreach (or in email), make sure they have analytics built in. Bitly (*https://bitly.com/*) is a great tool for this: it can tell you what links are getting clicked and where they are being reshared.

Misfit Wearables found that tracking inbound interest helped establish the right time to reach out to media partners with story updates. It also paid close attention to its perk data: what was resonating versus what was flopping, which colors were the most popular, and what kind of premium people seemed willing to pay for "limited edition" versions. You can monitor engagement around each reward type in the dashboards of Indiegogo and Kickstarter, among others.

There's also a constantly growing list of startups that offer tools to supplement crowdfunding raises. BackerKit (*https://backerkit.com/*), for example, helps with customer relationship management. The Kickstarter Status Board plug-in for Chrome is a clearly organized real-time project dashboard. Kicktraq (*http://www.kicktraq.com/*) calls itself "Google Analytics for Kickstarter." It offers algorithmic prediction of success or failure.

Academics are paying attention to crowdfunding as well. A team of researchers in Switzerland recently launched a site called Sidekick (*http://sidekick.epfl.ch/*), which predicts the likelihood that a campaign will succeed. In October 2013, they published "Launch Hard or Go Home!" (*http://bit.ly/predict_kickstrtr_success*), a paper describing a model they claimed could predict the success or failure of a crowdfunding campaign within four hours of its launch with 76 percent accuracy. They trained their model using Kickstarter campaigns from September 2012 through May 2013. They crawled these 16,042 raises, monitoring each campaign's progress using both a time series analysis of dollars raised on Kickstarter's site and mentions of the project on Twitter (social currency). The researchers also crawled the backer list for each project. Using predictive modeling techniques, they considered the rate of funds coming into a campaign. Following the presentation of the paper, they launched Sidekick.

It's worth doing some Googling to look into the latest tools available when you're about to start your crowdfunding raise.

PUBLISHING UPDATES FOR YOUR COMMUNITY

While waiting for your dollars to roll in, publish updates that keep your new community up-to-date with your progress. You want to engage these early adopters and turn them into evangelists, particularly while the raise is live! Their emotional investment is a valuable source of traffic to your campaign while it's running, and it will help you grow your company once it's over.

Consider posting update videos showing off development progress, a new version of your prototype, your office space, etc. Post links to news coverage. Highlight fundraising milestones ("We reached 50 percent of our goal, thanks to you!") or new perk offerings. Announce a new stretch goal.

John from Kickstarter advises using updates as a call to action to energize your base, and suggests having a strategy planned out in advance. Some potential calls to action include asking backers to help you spread the word for one final promotional push as a campaign winds down. Some founders offer special perks to backers who help drive sales through referrals. Others reach out with a survey or ask for feedback on a product idea.

Danae from Indiegogo adds:

> One of the biggest mistakes we see is people using Indiegogo solely as a way to ask for money and not as an R&D platform. They're not maximizing the opportunity they have to invite their community in, to engage with them, get feedback. This is a time when people are voting with their dollars. It's real feedback, way better than any you'll get from a focus group or Like button. Use your raise as an opportunity to get smarter!

Growing a community can be incredibly time-consuming. You'll want to be both *proactive* (sending out updates) and *reactive* (responding promptly to emails, particularly if they're about concerns). It's important to make the community yours: consider inviting them to sign up for a mailing list or directing them to a community blog that you'll continue to update after the project has shipped.

Updating should continue even after the raise is done. Keep your backers in the loop about your fulfillment process.

Beyond Crowdfunding: Fundraising for a Company

While many first-time founders think of crowdfunding as a way to raise money for their company, the real value of these platforms is that they can help you research a market, establish demand, and build a community. The money that you're raising is for a specifically defined *project*: that is, something with a clearly stated purpose, a start date, and a delivery date.

Crowdfunding is primarily used for working capital. And while you're building margin into all of your reward tiers (thereby hopefully turning a profit from this first run), even the excess funds that you raise will likely not be enough to turn your project into a company with enough money for salaries, marketing, R&D, and other expenses.

Scaling a hardware startup is capital-intensive. Chapter 9 covers the options for fundraising that can help you grow from a crowdfunded or bootstrapped project into a full-fledged company.

CHAPTER 9

Fundraising

HARDWARE STARTUPS ARE A CAPITAL-INTENSIVE UNDERTAKING. AS YOU'LL see in Chapter 10, most crowdfunding raises are not a path to funding a company. They provide an infusion of cash that will enable an entrepreneur to work full-time on a product and hopefully take it to market, but a majority of the funds brought in are going to flow back out to pay for production costs.

However, now is a great time to fundraise as a hardware entrepreneur, because there is an increasingly large pool of different types of investors to draw on. This chapter covers how to identify and form relationships with the right investment partners, how to articulate and sell your vision, and how to manage and streamline the fundraising process.

Not everyone wants to turn their project into a big business; it's important to have a clear idea of what kind of company you'd like to grow into before starting down the path of raising capital. If you're reading this book primarily to learn how to bring a single product to market through direct distribution, you can probably skip this chapter altogether.

There are steps you can take to achieve profitability without any external cash. For example, if you've hung out your own shingle on a platform like Shopify, you can continue accepting orders postcrowdfunding, or perhaps your crowdfunding campaign generated enough press to attract the attention of retailers right away. But most likely, at some point you will have to seek outside investment. To grow a hardware company, you need access to funds for salaries, paying your factory, marketing, warehousing, fulfillment, customer support, and more.

If you're looking to be a small ("lifestyle") business, raising institutional venture capital is probably not advisable; you might find yourself butting heads with investors who push you to grow quickly and expand into other offerings. A small business, particularly one with a unique or niche product, can grow slowly and steadily by reinvesting profits or taking a small investment from a supplier or partner. If you're looking to compete with an established or large incumbent, however, you'll need capital not only to produce your physical goods, but also to build a brand. You'll be spending money on marketing and might need a sales force, retail partnerships, and more. And if you're looking to build a large company with multiple products, you're going to need research and development funds to design and test new products while continuing to manufacture and sell the others.

First Things First

If you're trying to decide between participating in an accelerator (discussed in Chapter 7), creating a crowdfunding campaign (Chapter 8), or raising a round, it can be difficult to know what to do first. The biggest factor in deciding how to proceed is the background of the founding team. Do a critical self-assessment and ask yourself whether your prior experience as an entrepreneur makes you a good candidate for "raising on a dream." If you have no prior experience starting a company or working with hardware, it is typically difficult to sell investors on your vision alone; you are unlikely to successfully raise a funding round prior to a crowdfunding raise or participation in an accelerator. You'll need to find a way to inspire confidence in your ability to execute.

Even if you have a compelling background, larger investors (venture capitalists, strategic partners, or "super angels" writing six-figure checks) typically want to see evidence of market demand for a product. If you haven't sold anything yet, a successful crowdfunding campaign can help prove that market demand exists, and might make a capital raise easier. If the project is a runaway success, you'll likely find that investors will be reaching out to you.

That said, if you are a previously successful founder or hardware expert, it might be worth your time to try to seek investment *before* the crowdfunding campaign. Cash in the bank can provide a cushion for a production run, new hires, or additional marketing, and it's one less thing to worry about as you set about fulfilling orders after a successful crowdfunding campaign. You should gain a sense of how interested investors are with a handful of meetings across a few weeks, and if concerns about demand or team come up repeatedly, you can proceed with an alternative course of action.

Many first-time hardware entrepreneurs find that participation in an accelerator helps open doors that lead to checks, because the admission process provides validation, and the three to six months of supported and mentored development can help founders execute more efficiently.

Bootstrapping, Debt, and Grants

Bootstrapping is the term for launching a self-financed company. The founders still incorporate, but rather than accepting investments from outsiders, they use their personal savings to fund early development. This is how most people fund their MPVs, because it's pretty tough to raise outside cash for an idea on a napkin.

Bootstrapping a hardware startup can be extremely challenging, due to the capital- and time-intensive nature of building a physical product. It's difficult to iterate cheaply at any point in the hardware development cycle. Manufacturing problems, design flaws, wasted materials, and the risk of recall are particularly stressful when cash is in short supply. At the same time, the design or prototype stage is also the period during which funding from angel or institutional investors is hardest to come by. As a result, many entrepreneurs fund their startup by drawing on their own savings.

Others choose to take on debt, particularly if they believe that there is clear path to revenue. Since many banks will not write loans for startups, this often takes the form of credit card debt or a personal loan. Some US-based founders who commit a certain amount of personal capital to their new company might qualify for a Small Business Association (SBA) loan. The SBA.gov (*http://www.sba.gov/*) site offers several different programs, so it's worth a look to see if you meet the various criteria. The Loan and Grant search tool (*http://bit.ly/loan_n_grant_search*) might help you uncover programs that can help you. However, investors often view debt on the books unfavorably if there's a need to raise a future equity round, so it's something to avoid unless you are certain it will bring your company to a point where offsetting revenue can pay it down relatively quickly.

The federally funded Small Business Innovation Research (*http://www.sbir.gov*) (SBIR) program offers another option: grants for small companies based in the US. The purpose of the program is to support early-stage technological innovation or research and development work.

Well-known tech companies such as Symantec, Qualcomm, and iRobot received early SBIR funding.

The SBIR program consists of three stages, which are broken down according to criteria designed to gauge the technological feasibility of an idea and the progress of the company over time (see "How to Navigate the SBIR Process" on page 193 for Todd Huffman's tips on navigating the process):

Phase I
> Typically a feasibility study, the first phase is designed to verify that an idea or new technology is feasible and has commercial potential, and that the team is capable of executing well. A Phase I grant traditionally provides up to $150,000 over a six-month period, though the exact amount may vary according to the budget of the specific federal agency backing it.

Phase II
> In this phase, the technology and approach vetted in Phase I is further refined and a prototype is developed. A Phase II award is contingent upon meeting success metrics during Phase I. It has a funding cap of $1 million over a period of up to two years.

Phase III
> This phase is the commercialization period. It isn't funded by further SBIR money, but participants are often eligible for other federal funding programs, and occasionally for direct contracts with the US government. It is possible to go right from Phase I to Phase III.

In addition to SBIR, the Small Business Technology Transfer (STTR, also at *http://www.sbir.gov*) program provides similar grants. These are designed to facilitate public/private partnerships. The company must maintain a formal collaboration with a nonprofit research institution or university located in the US. Applying for a grants might be time-consuming, but it's a great way to raise early funds without giving up any ownership stake in your company. Before attempting the application, read one of the many blogs—such as SBIR Coach (*http://www.sbircoach.com/*) and SBIR-STTR Grants Help (*http://www.sbirsttrgrantshelp.com/*)—dedicated to strategies for writing a project proposal.

How to Navigate the SBIR Process

Todd Huffman is the founder of 3Scan, which aims to automate pathology with its Knife Edge Scanning Microscope (KESM). He's also the Vice President of IST Research, which builds technology for austere environments and postconflict situations. In those roles, he's successfully applied for several SBIR grants. Here he shares some tips on the process.

The first thing to do when looking into the SBIR program is to visit the SBIR Gateway (*http://www.zyn.com/sbir/*). The topics are organized by agency and include software, hardware, life sciences, and more. Solicitations are released in phases, and the Gateway is both a search engine for all SBIRs that are open at any given moment in time and a calendar that lists agency solicitation release schedules.

Each solicitation specifies the name of the program manager. You can also see past requests, which can help you get a sense of which government agencies might be potential clients. Todd explains, "It's instructive to look at the history of a program manager or agency's SBIRs, because you can see the technology arc that they're interested in." The Gateway also offers a mailing list, providing project solicitations to interested parties as they are released.

So how can you make your application more likely to succeed? The most important factor is that you must understand the needs of the program manager. Program managers are selected based on their operational experience, and they're putting out the request because they have a real problem to solve and haven't found a solution within the existing commercial market.

Understanding the problem lets you reach out with intelligent questions and determine whether what you're building is actually a good fit. "There's a period of time in which you can email the program manager questions," Todd says. "They'll usually respond. It's absolutely reasonable to ask them, *Is this the kind of thing you're interested in?* Even if you don't get a response, it's good to plant the seed of an idea in their head." Just don't ask, "Will you fund this?" You're trying to discover their pain point, not pitch your startup.

First-time applicants often don't realize that the SBIR structure is somewhat flexible, and that program managers have a degree of discretion. IST Research submitted an application for a Navy SBIR requesting Android smartphone applications for epidemiological analysis. Although Phase I is typically a feasibility study only, the team offered to do the study, and—in exchange for $50,000 in extra funding—to build a functional prototype and demo it as well.

In the same submission, the team also suggested an alternate approach to the specified request: although the proposal called for smartphone-specific apps, they suggested a "dumbphone" approach might be more relevant for solving the underlying problem. The program manager (a Navy doctor) accepted the offer and the suggestion, and they received the Phase I funding.

This kind of interactive process can help a startup continue to keep moving rapidly despite the traditionally slower pace of government. In some cases, such as with DARPA, it is occasionally possible to apply directly to Phase II, provided that you can demonstrate the equivalent of a Phase I feasibility study.

One of the benefits of the SBIR program is that agencies can pick up projects from one another. As the project for epidemiological analysis apps progressed, the Navy doctor chose to move ahead with smartphones only for Phase II, so IST Research did not continue with the project. However, DARPA later approached the team requesting similar technology. DARPA issued a Phase II solicitation, picked up the team's Phase I work, and provided them with a million dollars in funding. Soon after, the Naval School had a similar problem and issued a Phase II and III call simultaneously, giving the team another half-million dollars and an open-ended contract to market their work to the rest of the government.

Todd emphasizes the fact that the SBIR process is one of the easiest ways for a startup to get a government contract:

If the government is a likely customer, SBIRs are really great. This is because if you're a small company and someone in the government wants to use your product, they can't just buy it. They have to go through a government contract process, and the small company has to be compliant with all of the federal contracting rules, which is onerous. If you have a Phase III SBIR with one agency, you already have a contract vehicle and can sell to other agencies without having to go through the full contracting process. The program managers are also used to dealing with rookies and are quite patient.

Occasionally, the manager who created a specific Phase I or Phase II request will make introductions to other agencies that she thinks would be interested when it's time for Phase III.

The SBIR/STTR program does have certain weaknesses. "The bureaucratic overhead can be a bit much for small companies," Todd says. "The process is meant to be simple, and the program managers are responsive, but it can take a while." From start to finish, Todd has found that it takes approximately five to six months from submitting a successful application to receiving funds.

> Still, he emphasizes that SBIR is worth considering for hardware companies in the B2B or B2G (business-to-government) space: "It's nondilutive funding. It's not going to be something that funds your whole company, but it can certainly help quite a bit." That said, if the government isn't one of your big potential customers, or if your product doesn't fall within the scope of the problem the SBIR brief outlines, it's probably not worth your time.
>
> So, to recap: Phase I is to help you flesh out your idea. Phase II provides funds to help you build it, and Phase III is a vehicle for you to sell it.
>
> If you're interested in learning more about the application process or how to write the best possible proposals, look for university courses taught by experts who can guide you through the process.

In addition to the federal programs discussed above, many states and large cities have their own Economic Development Councils (EDCs), which offer support to local entrepreneurs. New York City's NYEDC, for example, offers funding programs, business plan competitions, mini-incubators, and free classes designed to help new entrepreneurs launch small businesses. They also list resources, such as real estate tax deductions and business incentive rates from utilities, that can help bootstrapping entrepreneurs run lean.

If you are starting your company outside of the US, or have a non-US founding team, investigate similar programs and resources in your country. Canada, Chile, Singapore, and Brazil, for example, offer a mix of grants, funding, visa assistance, and tax breaks to startups.

Friends and Family

If you're tapped out on bootstrapping and ineligible for any grants, the most friendly faces you can reach out to for money are your friends and family. They know you and trust you, and you've presumably already proven to them that you're reliable and can do the things you put your mind to. They've probably heard you talk excitedly about your idea.

However, most people find it difficult to raise more than a small amount of funding this way. Typically, they're able to round up an amount in the low six figures at the most. Tapping your personal network has some downsides as well. A small check ($10,000 to $25,000) in the business world might be a large check to the family member who wrote it. If he's never invested in a company before, it's important to make sure he understands how risky an idea-stage investment is, and that he runs a very real chance of losing the money he's putting into your company.

$100,000 is a drop in the bucket if you're building something that requires major production work, or specialized warehousing or logistics. However, if you're a first-time entrepreneur looking for funds to build a prototype before getting to that next stage, your best bet is likely a combination of self-funding (bootstrapping) and raising money from friends and family.

Angel Investors

Angel investors are individuals who invest their own personal capital in early-stage startups. Sometimes a group of individuals invest as part of an *angel syndicate*, but often they are simply independent investors who have a tech or startup background, have had a successful exit, and/or are independently wealthy.

They typically write checks from $25,000 to $100,000, although some (who are colloquially called *super angels*) occasionally go up into the $250,000 to $1 million range. Angels often invest in areas in which they have personal expertise or an extensive network. They can be "just a check," but many choose to be hands-on investors who actively help their companies grow.

The number of angel investors and angel syndicates has risen quickly over the last 10 years. In 2002, there were approximately 200,000 active angel investors and approximately $15.7 billion in investment dollars. According to the most recent angel market analysis report, which was in 2013, the number of active angels is 298,800. Capital invested grew to $24.8 billion across 70,730 ventures. This was an increase of 5.5 percent over 2012. The average size of an angel deal in 2013 was $350,830; average equity received for investment was 12.5 percent; and average deal valuation was $2.8 million. Many angels invest through networks such as Golden Seeds and Tech Coast Angels. The annual Halo Report (*http://bit.ly/halo_report*) tracks which groups are the most active and can be an excellent resource for identifying investors in your sector or regions.

Because startup investments are considered a risky asset class, SEC regulations require that angel investors meet the legal accreditation standards (*http://www.sec.gov/answers/accred.htm*) for individuals. Currently, an individual investor can meet those standards in one of two ways: either by having a net worth exceeding $1 million (not including the value of her primary residence), or by having income exceeding $200,000 (or $300,000 if combined with a spouse) in each of the two most recent years and a "reasonable expectation" of the same income level in the current year.

THE JOBS ACT

A recent piece of legislation, the JOBS (Jumpstart Our Business Startups) Act, aims to change the requirement that investors in nonpublic companies be accredited. Under this legislation, nonaccredited individuals will be able to invest in new "emerging growth" ventures via government-registered *funding portals* called *equity crowdfunding sites*. These are different from project-specific donation crowdfunding platforms such as Kickstarter, because the investments are made in exchange for equity rather than "rewards" or preordered merchandise.

Previously, a private company could have only 500 (accredited) shareholders on its books before it was required to meet SEC public reporting and disclosure requirements. The JOBS Act raised that to 2,000, of whom 500 can be nonaccredited. There are caps placed on nonaccredited persons investing via crowdfunding sites: the greater of $2,000 or 5 percent of income for people earning up to $100,000 a year, or the lesser of 10 percent or $100,000 for people earning above $100,000/year. Companies that choose to fundraise in this way will also be responsible for taking "reasonable steps" to verify the status of their investors.

 Although the bill was passed in April 2012, at the time of this writing, the SEC is still finalizing the structure of these new rules.

ANGELLIST

For founders who are new to entrepreneurship, finding and connecting with angel investors can seem daunting. One of the best resources for plugging into the angel community is AngelList (*http://angel.co*). Started in 2010 by Naval Ravikant and Babak Nivi, the site has grown from an email list (the origin of the name) to a comprehensive platform for enabling entrepreneurs to connect with both angels and early institutional investors. As AngelList's popularity has continued to grow, it has become one of the first stops that an investor makes when looking into a potential investment or getting a sense of what's happening in a given sector or region.

The site is designed as a network, and it enables investors and founders alike to showcase themselves. Founders can create rich company profiles that feature information such as a product video or slide deck; press coverage; traction information; incubators they've participated in; quotes from advisors, investors, and customers; and more. Investors also create profiles, tagging themselves with the sectors they invest in and the check sizes they write, and linking themselves to their portfolio company pages.

AngelList emails the profiles of suggested companies to targeted investors, who can reach out and get an introduction to the founders. Users can also browse one another's profiles and follow one another's activity, discover trending startups, source talent or support staff (e.g., attorneys), and much more. It is a thriving community and a valuable resource.

It's important to create a polished presence for your company. A top-notch profile can attract the attention of the AngelList team, which may feature you. With enough attention, you may become a *trending startup*. In either of these cases, your startup will be emailed out to hundreds of investors and potentially showcased on the site's front page.

In anticipation of the JOBS Act rules being finalized (see "The JOBS Act" on page 197), AngelList has recently made it possible to raise money directly on its platform. Startups that have a *lead angel* (someone who has committed a minimum of $100,000) are eligible to close out the rest of their round via a *self-syndicate*: they post the raise to the platform itself.

Angel investors can form syndicates as well. They commit to investing a certain amount of capital in a specific number of deals per year. Other accredited investors ("backers") can join a syndicate, committing to invest alongside the lead. This lets founders potentially receive a much larger investment through a connection with a single angel. For more information about how syndicates work, see AngelList's Help page (*http://bit.ly/syndicates_angellist*).

In "AngelList How-Tos" on page 198, AngelList product manager Ash Fontana discusses how to make the best use of the platform.

AngelList How-Tos

One of the most daunting things for many new entrepreneurs is the prospect of reaching out to investors and establishing a network ahead of a raise. Since AngelList's launch in 2010, the team has worked hard to make that process easier. Ash Fontana, product manager, shared a few tips on how best to leverage the platform.

Using AngelList properly means being willing to put yourself out there, in several ways:

Give the market as much information as possible.

Create a detailed profile and think about how to make it the best possible public representation of your startup, almost like a landing page. Use the tagging system to make sure you will show up in searches by industry sector or geographical region. For hardware products in particular, add lots of images of people using the product to your profile. If you have units already in production, add some video! Videos help convey a sense that the product is real. AngelList proactively features the best profiles on the site, emailing them out to investors. In order to be considered, an information-rich company page is a must.

Use AngelList like a social network.

That's how it's architected. This requires an investment of time on your part. Just as you build up LinkedIn and Twitter contacts and relationships over time, you have to put some effort into building connections on AngelList. Check in daily, and monitor your feed to see what people are doing or investing in. Add investors and advisors to your company profile as you bring them on. Update with important news, and have friends and followers share the updates so that you appear in their connections' newsfeeds.

Be proactive!

It's natural to feel somewhat nervous about reaching out to investors, whether in person or online. It's important to remember that they're on AngelList because they are looking to fund companies. So get out there: request introductions, follow founders and investors, and message all of the people you can message...provided, of course, that you have reason to believe that they're interested in your space (investors also tag their profiles with their interests).

The time to join AngelList isn't just before you want to start fundraising; it's when you're comfortable announcing yourself as a company. So get on there, build a great profile, and start forming relationships as soon as you're ready to publicly acknowledge that you're a company. Then, when you're ready to fundraise, flip on the Fundraising switch on your profile.

Once you're actively raising, you might want to update your profile with information specifically geared to investor questions. Add some details about your manufacturing process and relationships if you're raising more than a small seed round; investors want to know that you're on top of this. Share presale numbers, signs of traction, and established customer relationships.

The typical process of successfully raising a round involves reaching out to many investors. The best way to do that is to be authentic. Keep a first note simple, and be sure to include exactly why you think that specific investor is a fit for your particular startup.

Once you've made a connection with an investor who decides to write you a check, that signal often leads others to follow. One of AngelList's offerings, Invest Online, helps to facilitate this on the platform itself. Companies that have a demonstrated commitment of $100,000 from an AngelList investor are eligible to fundraise online. Companies participating in the program are emailed out to the broader AngelList community. It's a great way to find sources of capital outside of your network, and it generally takes less outreach effort.

AngelList Syndicates are another, relatively new, way to fill out a round. Well-known angels and seed investors form syndicates on the platform—minifunds, in a sense, seeded with their own capital but filled out with money from smaller investors. This means that an individual angel who ordinarily would have written a $25,000 check has a bigger group behind him and can conceivably now contribute a much bigger amount, in the hundreds of thousands of dollars. Reaching out to these investors is a bit like approaching a fund.

More than a third of the investors on the AngelList platform are institutional investors, from VC and seed funds. It's more than a place to find your first $100,000. And as a startup itself, AngelList is constantly launching new features to facilitate connections and access to capital. Be sure to check out the AngelList blog for the latest tools to help you get out there and raise money.

Hardware is a particularly popular category on AngelList, and it attracts a lot of investor interest. "It's always good to get some presales on Kickstarter or elsewhere, and then go on AngelList to raise an equity round," Ash says. "This is a common and successful strategy; investors see it as validated demand. But ultimately, at the angel stage, it's about the people and the product."

Venture Capital

Venture capitalists are professional investors who provide capital to young, high-growth-potential companies. A majority of the capital comes from outside investors (*limited partners*, or LPs) and is pooled in an investment vehicle called a fund. Because the life cycle of a fund is typically 10 years, venture capitalists target investments that will achieve a return for them within this time frame. Generally speaking, they expect to lose or break even on most of the investments in a given fund, but earn outsized returns on a few.

Raising a venture capital round is hard work, particularly for a hardware company. Historically, many venture investors have avoided hardware because of the high cost to bring a product to market, difficulty of rapid iteration, and challenge of vetting market demand prior to product release.

As discussed earlier in this book, those concerns have been somewhat mitigated recently, and an increasing number of institutional investors are putting money into hardware startups. It's difficult to get exact stats on the flow of money into hardware startups, but data from DJX VentureSource (*http://bit.ly/hardware_trends*) indicates an increasing amount of investment dollars in the sector: in 2012, $442 million was invested into hardware startups. By 2013, that number had nearly doubled, to $848 million.

Even though more checks are being written, fundraising can still be a long slog. Many entrepreneurs on both the hardware and software side will tell you that fundraising becomes their full-time job for several months, until they manage to close a round. Since you'd probably like to minimize that phase and get back to product-building, let's go over some best practices for successfully closing deals with institutional investors.

TARGETING INVESTORS

It's difficult to overemphasize the need to choose your target investors carefully. It's important to know what value-add you'd like your ideal investor to provide.

A venture investment is a long-term partnership, so it's important to find people who will be able to help you grow. A first venture capital raise should be used to help you scale and find product-market fit. In the short term, you need partners who can help you reach the milestones you're raising money to hit. If you haven't yet gotten to market, for example, you might want an investor who's helped other portfolio companies navigate the manufacturing process. If you're selling something into a niche channel (say, a health tech device), you might want to find investors who can help you with connections into hospitals, or who have navigated an FDA approval process. Money is money; the value-add of investors is in the extent to which they can help you grow your company.

The fundraising process is similar to sales: you're selling a vision. Start by building out a fundraising pipeline. Identify a set of investors you want to target and set up a spreadsheet or customer relationship management (CRM) system to keep track of contact dates, feedback, and requests for follow-ups or additional information.

How do you identify investors and populate the pipeline? Do your research and be selective. To find the investors for you, read news articles, industry publications, and blogs focused on your sector to discover who the active participants are. Check AngelList's investor profiles; most VCs on the platform have tagged themselves with relevant sectors or geographic areas that fit their investment theses.

CrunchBase (*http://www.crunchbase.com*) and Quora (*http://www.quora.com*) are also good resources. CrunchBase releases monthly database dumps as Excel files. Get in there and sort by sector and date, and see what funding events are happening in hardware and who's participating. If you'd like to hear more about an investor's process or what she is like to work with, reach out to connections in the portfolio.

Not all venture capital funds are the same. Some don't invest in hardware at all. Some, especially smaller funds, won't touch certain sectors with large up-front capital requirements or significant risks (e.g., healthcare devices requiring FDA approval).

Be aware that most VCs won't fund companies that are potentially competitive with their existing investments, because this can create conflicts of interest. Fortunately, most VC firms' websites include a portfolio page, so you can get a sense of what the firm looks for.

PERSONALIZED INTRODUCTIONS

After you've identified the investors who are a best fit for your company, it's time to reach out for a meeting. The ideal way to do that is to find a mutual connection who can send a warm introduction. LinkedIn is a great place to discover how specific investors are connected to you through your personal network. Entrepreneur friends who have founded or work at one of the VC's portfolio companies are another great way in.

If you don't have a strong network yet, don't be discouraged. Creating your own warm relationships with investors is much easier than many founders think. If you have a particular reason to want to raise from someone, you probably are interested in his advice just as much as money. So, ask for it. Reach out over email or Twitter. Ask for a coffee or a quick call, and specify that you'd like to hear his take on a problem or space. It's important to be specific in your ask. Just saying you'd "like to talk" leaves the investor wondering if you're beating around the bush about fundraising.

It's a great idea to do this several months before a raise, so that you can solicit feedback and then demonstrate progress over time. Mark Suster has an excellent post about establishing relationships on his blog, Both Sides of the Table, where his advice is to Invest in Lines, not Dots (http://bit.ly/invest_in_lines). A "dot" is a single interaction—your company is at a distinct place at that moment. As a relationship develops, there are more meetings, calls, or email updates—more dots. Eventually, the investor can draw a line connecting these dots, see the path the company has taken to date, and predict where it might go in the future. The truth is, it's rare for an investor to write a check to a founding team that she has just met, or that comes with no validation through a personal connection.

Sometimes, it's easiest to establish a relationship with an institutional investor who isn't a partner at the fund. Analysts and associates might not have check-writing power, but it's their job to meet interesting entrepreneurs, and they are more likely to have time for a coffee and feedback session than a partner might be.

While some Silicon Valley conventional wisdom will tell you not to waste your time with junior investors, their incentives are aligned with yours. They want to bring good companies in to pitch the partnership as much as you want to be in there pitching. Associates will often help guide a founder through the process and advocate for their favorite startups in the firm's weekly meeting.

It might mean a few extra meetings before you get to the partners, though, so if you're truly short on time or about to close a round, it isn't rude to make that clear.

While there are many approaches to establishing a relationship, sending a 30-page deck to the blind-submissions email address (e.g., *plans@vcfirm.com*) is not one of them. Those email addresses are rarely checked, because investors receive hundreds of emails to their "real" email accounts on any given day.

If you have no connections at all to an investor whom you want to talk to (and no way to make them), try to at least discover his direct contact information. Many institutional investors are also on AngelList, so you can reach out to them via the platform's Messages service.

TELLING A STORY

Once you've landed the meeting, it's time to tell your story. A good pitch is a story about a problem and a solution. It's a narrative that weaves together both your existing progress and your future vision. If you have a prototype or demo, craft the flow of your story around showing it off. Showing is always better than telling.

The best pitches touch on the following things:

The problem

What is the customer pain point you're trying to solve? Be specific. What is it about the problem that has appealed to you on such a deep level that you're willing to devote years of your life to solving it?

Your solution

What is your fix for the pain point you've just articulated to the investor? The solution doesn't have to be absolutely innovative; it's fine to be building a better mousetrap, as long as you're able to articulate specifically why your mousetrap is better. Are you competing on price, on features, or on something else entirely? (Hint: you identified your differentiators in Chapter 3!)

Your team

Why is your team the best possible group of people to be solving this problem? Do you have a background or personal experience in the space? The founding team is the most important criterion for many VCs. Ideas often change, so investors back solid people in who inspire confidence.

If you have previously built a company or successfully launched a product, be sure to emphasize that. Even if it's not relevant to the specific space you're working in now, VCs like to back founders who have demonstrated an ability to execute. If you're a first-time founder, focus on professional accomplishments and the milestones you've already hit on the product you're pitching.

The addressable market

Who are the people suffering from the problem that your widget is solving? How many of them are out there? What percentage is likely to pay for your product? Many founders find this difficult to quantify, but it's important to understand the economics of the space you want to sell into.

You have to be able to make a compelling case for why your market is big enough to be appealing to an investor. Do your homework here, and be honest. There is a difference between *total* market (e.g., all of the teachers in the country) and *addressable* market (e.g., the teachers working in school districts with a budget capable of buying your awesome new ed-tech device). (We discussed this in terms of TAM/SAM/SOM in "Market Size" on page 37.)

There are many products out there, hardware or software, that are beautifully designed and in demand by some group of people. But if there aren't a lot of potential buyers as well (the SOM is small), VCs are going to be hesitant to back your company. You might hear the term "lifestyle business," or "not venture-fundable," in their feedback to you. This means that your idea is good enough that some people will undoubtedly buy it, and you might earn a nice living from it, but it's probably not going to grow into a billion-dollar company capable of generating the returns that venture investors are looking for.

You might be asked, "Are you a product or a platform?" This is the investor's way of asking if he's going to be backing the one widget you are currently producing, or if there is a vision for a stable of products. Will you be content making a single connected can opener, or are you striving to be the next Oxo? This is of particular concern for startups that don't have a software component.

It can be difficult to build community engagement and brand loyalty around a single simple physical product. If that's all you're personally interested in making, that's fine. But it's going to be difficult to convince an institutional investor that a product line consisting of one type of mousetrap is going to sell enough units to produce venture-caliber returns. A well-articulated game plan for how you will become the next Oxo is much more compelling.

Your traction and revenue

If you have any existing traction, get it out there, front and center. Investors love traction. If you are a hardware company with a software component, share your engagement numbers and relevant software metrics. Share preorder or sales data. Mailing list signup counts. The number of Kickstarter or Indiegogo backers you have, and the total dollars raised. Enterprise company letters of intent (LOI). Any data that you have that can indicate to an investor that there is demand for your product. Most early-stage investors aren't looking for some specific magic number of orders or amount of revenue; they care about trends over time.

The funding ask

Why are you asking this investor for $3 million? What does that $3 million get you? Be specific. Is it for hires, or marketing, or manufacturing? While most seed-stage investors ignore the financial projections of early-stage software companies, hardware is a different beast. Certain things need to be paid for: salaries, manufacturing, warehousing, shipping, etc. Don't forget about rent, legal costs, certification costs, and marketing.

It's important to have a clearly articulated estimate of what your costs will look like, particularly if you're asking for money to go to market for the first time. This doesn't have to be anything fancy, but it should convey how the dollar value that you're asking for gets you from point A to point B. The investor wants to see you asking for an amount of money that can plausibly give you 18 months of runway.

Your competition

Every startup has competition. Period. You might argue that your device is the very first of its kind to be conceived, but few problems are new, so your competition is whatever people are currently doing to overcome this difficulty. If they're doing nothing, then your competition is people not thinking this problem is important enough to bother spending money on your solution.

Be honest about your competition, because the investor is going to do her own competitive research as well. If you've conspicuously neglected to mention your closest competitor, you will seem either dishonest or uninformed about your market. Many founders like to draw a feature matrix, if they're competing on features, or a quadrant visualization (see Figure 9-1), if they're competing on multiple factors.

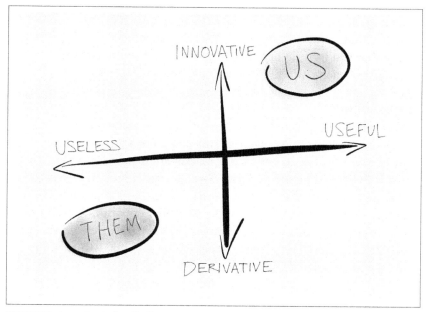

FIGURE 9-1. A quadrant visualization

Typically, each of the sections just discussed gets one, maybe two, slides. There are many great blog posts describing how to design an optimal slide deck. One of our favorites is Guy Kawasaki's "The 10/20/30 Rule of PowerPoint" (*http://bit.ly/102030_rule*) (10 slides, 20-minute presentation, no font smaller than 30 points). You can also hit Slideshare, where many startups share their presentations.

> **TIP** Do note that a deck that you send out is different from a deck that you present in person. You don't want to be reading your slides to the VC in an in-person meeting...and you don't want to send a minimalist deck with one word per slide to an investor over email.

If you're building something that is truly new, such as an innovative technology, be prepared for investors who don't understand what you're doing. VCs are great generalists, but you might find that few have had professional experience in your area.

Being able to convey why what you're doing is important to nonexperts is a valuable skill. You're selling your concept to the VC now, but you will have to sell the product to customers or explain it to the press eventually. Visuals or nontechnical diagrams in the deck are a great way to make things more comprehensible. An appendix is a great way to ensure that you have all of the information you need to answer the most common questions, while still adhering to the 10/20/30 rule.

DUE DILIGENCE

If your pitch goes well and the VC partnership is excited, most will begin the process of doing due diligence. This phase takes anywhere from a few days (if the VC already knows the market well) to a few weeks (if the investor wants to do some in-depth research or customer interviews).

The purpose of due diligence is to answer the "Who, What, When, Where, and Why" questions around a particular opportunity. At an early-stage fund, due diligence comprises market research, competitive research, and founder reference checks. The investor will examine the drivers underlying the market you're targeting. If you're working on a health-tech device, this will likely include the regulatory environment.

Competitive research typically takes the form of understanding who the entrenched big players are, and also doing a search for other startups in the space, to see how your offering compares. Founder reference checks will start with the names that you provide to the VC, but they will likely also include "back door" checks of people you and the investor have in common, or other names that your references provide. Customer validation phone calls are quite common, particularly if you're a B2B company. The VC will ask to see your cap table.

Finally, if you already have revenue and are doing a first venture raise later in the game, a thorough examination of financials will also be part of due diligence.

So, what makes a VC say yes or no? VCs most commonly do a deal because they love the team and they are excited by the problem. Building a product that solves a big, meaningful problem typically means that customers will be drawn to your solution organically (or at least that they will be willing to pay you for it).

Investors say no for a variety of reasons. Many health-related hardware companies find that investors are scared off by the 510(k) approval process. For consumer devices, the "product vs. platform" issue might make a startup seem unlikely to grow into a big enough company to tempt a VC. Sometimes, the pass is about your stage not aligning with where the investor prefers her investments to be (if this is the case, it makes sense to stay in touch occasionally and reach out again for a future round).

"No" is the most common response by far, so don't be discouraged if your first few investor meetings don't lead to checks. If you do get passed on, it's perfectly reasonable to follow up and ask for feedback as to why.

Strategics

A *strategic investor* is another type of professional investor worth a separate mention. Strategics typically represent the corporate venture capital or investment arm of a large corporation (e.g., Time Warner Investments or Dell Ventures). These funds generally invest because of the potential for a mutually beneficial relationship between the big company and the startup.

Sometimes, this takes the form of supporting a potentially complementary product or an experimental technology that might one day help the strategic investor enhance its own product line. There is often a financial motive: many of these firms specifically focus on areas in which they have deep domain knowledge, so they are likely to have a better-than-average chance at picking a winner. They will often coinvest alongside traditional VCs.

There are pros and cons to taking money from a corporate venture fund. The strategic partner might add value in the form of technical or business expertise in the industry. Some will offer access to or advice about marketing channels or management. A big brand publicly signaling interest might also provide a nice PR boost to a young company. Finally, some are simply acting as passive investors, who will provide a source of capital and are likely to follow on in later rounds.

However, on the flip side, there is typically higher personnel turnover in a corporate venture fund, so you might find yourself losing an internal ally and struggling to form a new relationship. Competitors of your strategic partner might be reluctant to become your customers. Occasionally, the strategic investor might require terms that specifically prevent you from selling your product in their competitors' distribution channels or retail stores. They might attempt to structure a deal more like a technology license transfer than a real equity investment.

There are also potential concerns about mergers and acquisitions further down the road. You might limit your pool of potential acquirers if one of your investors is a strategic competitor, because the acquiring firm might be hesitant to divulge information about the state of its own business.

A strategic investor with a large ownership stake might have the ability to block a sale. It is important to be sure that long-term strategic partnership intentions and incentives are aligned before accepting corporate venture capital.

Structuring Your Round

There are many excellent blogs and books that focus solely on the mechanics and intricacies of round structuring (*Venture Deals: Be Smarter Than Your Lawyer And Venture Capitalist* (Wiley) by Brad Feld and Jason Mendelson is a great one), so we'll be brief here.

When you go out to raise a round from angel or institutional investors, you have to decide if you're going to pursue *convertible debt* or *equity financing*. In a convertible note, an investor makes a debt investment (a loan) that converts to equity in the future, generally when a subsequent equity round has been raised. The note is typically structured with a set time period by which you must convert to equity or repay the loan.

A typical convertible note ask looks something like this: "We are raising $750,000 with a $6 million cap and a 20 percent discount." The first dollar amount is what the startup is looking to raise. There is often a conversion *trigger* that applies to future fundraises; for example, once $1 million in total has been raised, the debt will convert to equity. Otherwise, the note converts when a priced equity round is raised. The discount (20 percent in our example) is a reduction on the price of that theoretical next round, and it is designed to incentivize early investors.

If the company goes on to raise a subsequent Series A equity round at a $10 million valuation, the investor holding the convertible note receives her shares at a 20 percent discount. So if the stock price during that Series A round is on dollar a share, his earlier $750,000 debt investment converts to 935,700 shares of stock at 80 cents a share—a 20 percent discount in price. An investor entering during the A round would get 750,000 shares.

The other piece of the note structure, the *cap*, is a ceiling that limits the valuation at which the debt can convert to equity, and is designed to align entrepreneur and investor incentives and reward early investors. Consider the previous example: the cap is $6 million, and the valuation arrived at during the Series A negotiations is $10 million at one dollar a share. The convertible note holder's $750,000 investment converts as if the company were valued at $6 million: he gets 1,250,000 shares at 60 cents a share ($6 million/$10 million).

Generally, the debt will convert at the lower value of the per-share price as determined by the cap or the discount. In our example, the best option for investors is the 60-cent price from exercising the cap (not the 80-cent price via the discount). There are many possible intricacies in the terms of convertible notes, including interest rates and liquidation preferences, so make sure you understand what you're offering.

The main difference between a convertible note and a priced equity round is that no valuation is set during the former. Priced equity round math is a bit simpler. Rounds are quoted as "$X invested on $Y Pre" (e.g., $750,000 on a $6 million premoney valuation). The postmoney valuation, which is what the company is worth following the close of the equity raise, is determined by adding up the premoney valuation and the amount of investment. In our example, postmoney is $6.75 million.

Most institutional investors are looking to maximize their ownership percentage relative to the postmoney valuation. Typically, they will target between 10 and 20 percent ownership. Using our example, the VC's investment of $750,000 is 11.11 percent of the final $6.75 million company. Valuations are generally based on comparable companies. Historically, VCs have had a bit of an edge in negotiation here, because they see deals daily and are familiar with the closing prices of other deals.

However, industry transparency is increasing. Your banker can likely provide you with a list of deals similar to yours, and AngelList recently launched a Valuation tool (*https://angel.co/valuation*) that lets entrepreneurs check comparables themselves. Law firm Fenwick & West also offers an annual Seed Financing Survey (*http://www.fenwick.com/seedsurvey*), which can help you understand market trends.

Hardware companies are often more capital-intensive than software companies, so many choose to raise on a convertible note to avoid discussing valuation until there are more data points (e.g., sales). It is often faster to raise money via convertible note, because a rolling close is possible (i.e., not all capital has to come in at once), legal fees are lower, and you can avoid having to formally convene a board.

However, many venture capitalists feel that investor and entrepreneur incentives are misaligned during a convertible note raise, particularly if there is no cap. Many institutional investors also believe that their investment is more than just a check, and want to take a board seat and an active role in helping their portfolio companies grow. It is always best to approach an investor with an openness to negotiation, particularly if you believe she'll be a valuable partner in the long term.

If you're a growth-stage company (i.e., you've already raised a Series A, and you have a steady revenue stream), you might be eligible for a form of financing known as *venture debt* (*http://bit.ly/venture_debt*). Certain banks will offer venture-backed companies the option to take out a loan for working capital or financing a specific project, typically in exchange for interest and warrants.

The bank will occasionally also want the option to invest in a future equity financing round. If you're interested in using venture debt to fund a production run, there are helpful financial models (*http://bit.ly/venture_debt_blog*) online that can enable you to get a better sense of how the cash flows work, and whether it's a suitable option for your company.

CHAPTER 10

Going to Market

SO FAR, WE'VE COVERED HOW TO GET FROM VISION TO PHYSICAL PRODUCT and how to raise the funding necessary to seed your company. Now it's time to talk about building the business: business models, pricing, marketing channels, distribution, logistics, and more. Taking a product to market is a complex process. This chapter will focus on the most common issues you'll enounter on the road to getting your product into the hands of customers. Some questions that we'll touch on include:

What is your business model?

Many first-time hardware founders think their business model is obvious—they're going to sell a device! However, a number of other options might be more suitable in terms of both revenue and customer loyalty.

How much will your product cost?

Regardless of the business model you choose, you'll have to determine how much to ask your customers to pay for your device, or device experience.

Where will your product be sold?

Here we'll discuss the costs and the pros and cons of various distribution options. This will also be where we discuss raising awareness, advertising, and marketing, because strategies often vary depending on the distribution channel.

How will you fulfill customer orders?

Will you rent a warehouse and hire employees to ship the product, or employ the services of a third-party provider? We will examine the costs associated with various fulfillment options.

The goal of this chapter is to help you reach the markets and customers you've identified, in order to successfully sell your product.

Business Models for Hardware Startups

Let's start with identifying a business model. Hardware startups differ from software startups in that you are selling a physical *thing*, and most people expect to pay for physical things. In that sense, hardware companies have an advantage: they don't have to overcome the psychological hurdle that people often face when it comes to paying for software, apps, or content (given that so much is distributed for free). However, the obvious pay-for-the-product business model isn't always the best option for a hardware startup. Identifying a model that can provide you with recurring revenue is an even better choice, particularly if you're planning to launch additional devices or build out a platform and will have a time gap between product releases.

In Chapter 1, we segmented the hardware startup ecosystem into four parts: robotics, connected devices, personal sensors/wearable technology, and designed products. If you're making a simple designed product with no software or connected technology, your revenue is probably going to be derived from "vanilla" sales (pay-for-product), and you can skip ahead to "Pricing" on page 220.

For the other segments, this model might not be the best approach to building a profitable company. One reason for this is *defensibility*: as we've noted earlier in this book, the cost of building a device continues to drop. This is great for entrepreneurs, but it's also great for competitors who enter the market after you, or for unscrupulous folks who produce knockoffs (potentially a source of concern when dealing with overseas factories). If you are competing for market share and customer loyalty with the hardware alone, you may find yourself chasing ever-lower margins as these cheaper products enter the market.

The solution to this problem is to sell something in addition to the hardware—something more defensible. Ideally, you'll work to simultaneously build a brand reputation for high quality and offer a compelling value proposition (such as software with an exceptional user experience, or perhaps an engaged community) that keeps customers buying *your* device. Venture capitalist Brad Feld refers to the hardware products that he invests in as "software wrapped in plastic" (*http://bit.ly/feld_plastic*): the caliber and quality of the hardware is important, but the device is merely a piece of the overall product stack. Another VC, Chris Dixon, writes (*http://bit.ly/hardware_startups*), "Think of hardware as bringing the revenue and software/services as bringing the margin."

Consider the Kindle Fire and the iPad. Both are exceptional pieces of hardware. But more important, they are also vehicles for retailing content—apps, books, movies, music. A study by market industry analyst Vision Mobile highlights the differences (*http://bit.ly/vision_mobile*) among the iPad, the Kindle Fire, and Android tablet. Android tablets are not differentiated by a specific use case. They are "commodities almost from their inception, as the basis for competition among them is price." While the Kindle Fire is also based on the Android OS, it is positioned clearly and distinctly. It is a tool for media consumption—specifically, for consuming media purchased from Amazon. Amazon breaks even (*http://bit.ly/kindle_profit*) on the Kindle; it makes money on the content. As CEO Jeff Bezos put it, "We want to make money when people use our devices, not when they buy our devices." Although Apple also makes money from media sales, it makes the iPad device itself the core business and runs slightly above breakeven on content. People buy the iPad because of the exceptionally high quality of both the hardware and the user experience. Apple is a powerful lifestyle brand. If you do choose to make the hardware itself the focus, branding will be particularly critical to your success.

But more important than defensibility is the fact that a dynamic business model can help you continue to surprise and delight customers. If your product is a connected device or includes software, you can provide an evolving user experience by pushing new features, offering a subscription to tailored content, and more.

Here are a few examples of business models used by today's hardware companies.

SELLING ADDITIONAL (PHYSICAL) PRODUCTS

The first type of business model involves offering additional physical products that enhance your primary device.

Device plus supplementary (physical) accessories

One way to do this is accessories. Camera manufacturers do this with lenses, cases, tripods, and so forth. Phone-handset makers sell cases, earbuds, and other add-ons. Watch makers sell straps. If you have a product that lends itself to modularity, selling high-margin accessory kits lets users customize their experience to their needs while providing you with additional revenue.

Supplementary accessories can be a particularly appealing option for wearables companies, which can offer fashion accessories that appeal to individual styles or vary by season. Lumo BodyTech offers designer clips for its Lumo Lift product. Misfit Wearables sells a variety of necklace holsters and leather bands for the Shine fitness tracker. This model does require you to be able to produce the additional accessories, and you don't want accessory production to distract you from continuing development on your primary product. Some companies partner with other brands to reach new audiences; Misfit, for example, recently partnered with Swarovski to roll out glittery crystallized bracelets and necklaces to hold the Shine. Tory Burch designed a bracelet enclosure for Fitbit. Rolling out new add-ons can keep a base product feeling fresh while you're working on a new version. A partnership with another brand can help you reach a new demographic and expand your market share.

Device plus ancillary materials: razors and blades

Another variant of selling additional physical products is the *razors-and-blades* lock-in model, ostensibly named after safety-razor pioneer Gillette (*http://bit.ly/razors_n_blades*). Gillette's patent for its razor handle had expired, and its competitors were suddenly able to produce competing handles. In response, the story goes, Gillette elected to sell the razor at a loss, and profit on the blades. While you may not be in a position where selling the device at a loss is necessary, creating lock-in through compatible accessories is an option. 3D printer companies like MakerBot often generate revenue by selling high-margin filament, much like their paper-printer forebears who sold ink cartridges.

Device as platform for upselling other devices

This is the *platform play*. It's particularly useful for connected devices and wearables, but it can also apply to certain types of robotics companies. The idea is that if your products can communicate with each other, then buying two products from the same brand confers additional benefits beyond the basic features of each. In wearables, this might mean your bathroom scale communicates with your fitness tracker. In robotics, perhaps your vacuum robot communicates with your mop robot so that it runs immediately following the completion of vacuuming. Perhaps multiple robots coordinate on the completion of a particular task in such a way that purchasing two is materially better than only having one. In a connected home, products can interact with each other via M2M (machine-to-machine) communication mediated by sensors.

This business model is typically possible only after several devices have been built, which makes it difficult for hardware startups in the beginning (some choose to partner with other companies), but it's something a founding team might want to consider for a long-term roadmap. It's fairly common among connected-hardware products brought to market by large companies. One example is the Iris (*http://bit.ly/iris_strtr_kit*) connected home system from home-improvement store Lowe's. There are a handful of products available on the system, including motion sensors, contact sensors, a smart thermostat, and more. Lowe's packages them into kits targeting specific customer problems: the "Safe and Secure Kit," the "Comfort and Control Kit," etc. Lowe's also invests in connected-home startups as a strategic investor, often requiring that the startups make their products compatible with Iris. This enables Lowe's to continue to extend the platform.

SELLING SERVICES OR CONTENT

Another category of business models involves selling software, content, or services that support or run on the device. The era of connectivity has opened up a world of new business models for physical devices, as new features can be rolled out electronically, with a software push. These updates can keep a device feeling new.

Device plus subscription services

Subscription-service business models work best in support of products that meet an ongoing need. You likely have a modem in your home that provides you with Internet access; you probably didn't pay for the device, but you do pay a monthly fee to the provider. If you have a home security system, you might have paid for an installation package consisting of a certain number of alarms and other sensors. Or perhaps the alarm company gave it to you for free, or charged only an *activation fee*. Although the upfront cost may vary across companies, the providers make their money on the monthly fees for the related services. In his book *Free: The Future of a Radical Price* (Hachette), Chris Anderson walks through several scenarios in which companies use this model. In one example, he shows how

Comcast recoups the cost of a "free" DVR within 18 months by charging for installation and monthly subscriptions (often more than $10 to use the box) and upselling other services.

The challenge for a hardware startup considering this model is determining whether your users are willing to pay in perpetuity to access your functionality. In the case of a connected home alarm, this behavior is already the norm. But if you're envisioning a monthly charge to solve a problem currently fixed by a one-time-expense "dumb" thing—say, a dead bolt versus your connected lock—you might find it difficult to get customers on board with adding an additional monthly bill to their budgets. Nailing the user experience, providing value far beyond the one-time cost solution, and doing extensive price-testing is critical to successfully selling services. But if your device lends itself to this model, it's worth the effort. A services component is an excellent way to establish a long-term relationship with your customers.

Device plus digital content

Another model for software-enhanced hardware companies is the app-store model, mentioned earlier in the context of the iPad and Kindle Fire. You don't have to be a tablet or phone manufacturer to build a compelling app ecosystem. Sifteo, for example, is a company (recently acquired by 3D Robotics) that makes a connected toy called Sifteo Cubes, which comes with four preinstalled games. Users can download additional games at a cost of $8 to $12 per game.

Connected devices offer many new opportunities for *content plays*. Digital content can take the form of recipes for a connected kitchen device, workouts created by a celebrity trainer for a fitness wearable... there are so many possibilities. Analytics and usage data from connected products can help you deliberately target the content you're upselling.

In some cases, your team might be capable of developing a continuous stream of supplementary content in-house. However, scaling often requires getting your community involved: for example, you might foster an external developer ecosystem and sell supplementary software or content using a revenue-share model. Gaming startup Ouya is doing just that: manufacturing an inexpensive gaming console, offering a free development kit, and doing a 70/30 revenue split with developers.

Hardware devices with digital content can also be sold as gateways to communities, in which case a customer's purchase of the device can be considered the price of admission to a network. Buying an Xbox, for example, opens the door to a network of people to play with.

SELLING DATA

As data becomes increasingly valuable, selling it can be a lucrative source of revenue.

Sell device to consumers and data to third parties

This business model lends itself particularly well to sensor devices, which gather extensive amounts of data. A fitness-tracking device, for example, might be logging information about an individual's activity level, heart rate, and calories burned. This information is potentially useful to health services companies, such as insurers.

It's worth noting that *data plays* are rarely as lucrative as first-time founders expect them to be. "Data" is not necessarily synonymous with "meaningful information." Before banking on a data-sales business model, be sure that you've reached out to enough customers to verify that what you are gathering is something that they will actually pay for.

This model can be particularly tricky for a B2C device company to navigate, because consumers are sensitive about privacy and will react negatively if they discover that you're selling their information (trying to hide this fact in some interminable user agreement is a sure way to get bad press). If you pursue it, it's important to consult with a lawyer to be sure that the proper disclosures, data handling, and anonymization protocols are all in place.

Just sell the data

For some hardware companies, it might not be necessary to sell atoms at all. Your customers might simply need the data that your device is the best equipped to obtain. Understory, a hardware startup that builds weather sensors with sophisticated analytics capabilities, doesn't sell the hardware to its customers. Instead, it installs sensors in specific regions and gathers data, then sells it to cities, insurers, agriculture companies, and so on.

OPEN SOURCE

You might be considering an *open source hardware* (OSHW) business model. In an OSHW project, all necessary data about the hardware is available to the community for free: design specs, documentation, files, etc. Hardware released under an OSHW license *(http://bit.ly/ oshw_def_1_0)* specifies what portion of the design is open, and allows modifications and derived works "to be distributed under the same terms as the license of the original work." This lets the community participate in an ongoing development and revision process, which reduces time and costs. It also means that anyone can take the open design and produce the hardware. As of 2013 *(http://bit.ly/open_hardware_2013)*, the majority of these projects are electronics-related. Many are led by solo founders with backgrounds in engineering who work in their spare time and rely on community contributions to move the project forward.

There are several business models *(http://bit.ly/oshw_models)* commonly used in the OSHW space:

Dual licensing
   The software is available under both an open source and a commercial license. This is most often done if potential customers are likely to build on top of the open source code for a proprietary product. For more information about dual licensing, see Elena Blanco's post on OSS Watch *(http://bit.ly/dual_licensing_model)*.

Manufacturing to sell
   Having access to a factory is necessary to produce many types of open hardware devices. Simply being able to execute the design and sell the product is a business model by itself. In some cases—such as MakerBot, 3D Robotics, and OpenROV—a company will sell both a kit and an assembled version of the same product.

Technical support or services
   The hardware design is open, but customers pay for technical support or expert services.

While an open source model isn't for everyone, companies such as Arduino, Spark, and OpenROV are thriving. To learn more, visit the Open Source Hardware Association's website *(http://www.oshwa.org/)*.

Pricing

Regardless of the business model you decide on, identifying a price for your offering is one of the most important decisions you'll make. In his book *Pricing Strategy* (Cengage Learning), pricing expert Tim Smith states, "When thinking of prices, it is useful to consider price as the value that the firm captures in a mutually beneficial exchange with customers."

Your price sets the stage for whom you can reach, where you can sell, and how you will market. You'll want to develop your pricing strategy long before you manufacture your first run. At the latest, your price will come into play during your design phase; ideally it's on your mind from early on in the idea validation phase. As you're validating your vision with customers, you want to get a sense of how willing they are to pay for your solution to their problem. If you're targeting a particular demographic, research their purchasing power and spending habits.

As discussed in Chapter 2, *positioning* refers to the space in the market occupied by your brand. It's a relative concept: when we talk about price positioning, we are referring to the concept that the price of a product relative to its competition communicates something. Price is a component of brand positioning; the cost of an item has an impact on the customer demographic that purchases it.

A high price can signal that an item is high-quality, scarce, or exclusive. For example, the Birkin bag, made by fashion house Hermès, is a beautifully designed product. It is made by hand, features the finest leather and high-quality metal accents, and is released in limited quantities. The starting price for the bag is around $9,000, but rare and exotic materials or colors can push the cost into the hundreds of thousands of dollars. This price puts it out of reach for the majority of customers; it's an exclusive luxury good. The brand has a storied history (Mr. Hermès made saddles for royalty) and is a favorite of modern celebrities. At the end of the day, however, the Birkin is still a handbag, designed to enable its owner to transport small personal necessities from point A to point B. You can buy a handbag that performs that function equally well at Walmart for $20. With its marketing emphasis on low prices, Walmart is also using pricing strategy to solidify its own brand positioning as an extremely popular shopping destination for cost-conscious customers.

Your brand identity (discussed in Chapter 4) impacts the type of customers your product will attract. You can choose to be upmarket, middle-market, or mass market, so long as you remain consistent across price, market messaging, and sales channels. Once customer perception of your brand identity has been set, it's difficult to change. Large, established mass-market companies often purchase a distinct upmarket boutique brand rather than attempting to change the perception or positioning of their original product (think Ford Motor Company, which, at various points, purchased Jaguar, Aston Martin, and Land Rover).

So how do you decide where in the market to sit? Entire textbooks have been written on *strategic pricing* (tying your price to positioning) and *price optimization* (always having the right price for a given customer and distribution channel). If your product costs a fair amount to manufacture, even at scale, you are looking at a middle-to-upmarket product.

Pricing is difficult, but it's important to get it right the first time. You want to be sure that whatever price you choose will facilitate profitability. If you charge too little, it will be difficult or impossible to achieve profitability; it may even be impossible to complete manufacturing for your first run. Starting too low is also potentially problematic from a quality-perception standpoint. Think about the kind of quality you would expect from a $50 robot as opposed to a $5,000 robot. On the flip side, if you charge too much, you may alienate potential customers or open the door for a competitor that will take lower margins.

In behavioral finance, the concept of *anchoring* describes how an initial piece of information becomes the framework upon which subsequent decisions are evaluated. In the case of pricing, a customer's first exposure to your price will shape his view of what the fair price of the product should be. It's very difficult to raise prices for a specific audience once you've told them that a product costs $X. Consumers anchor to that price, and seeing it increase often makes them feel like they're getting overcharged. They tend to feel the pain of a price increase more acutely than they feel joy at a discount (this is an extension of the behavioral economics concept called *loss aversion*, researched by Daniel Kahneman and Amos Tversky). As a result, it is extremely challenging to raise your price without angering customers.

There are a few opportunities in the life of a hardware company for you to raise prices without negative customer associations. If you do a crowdfunding raise and discover that you didn't price in enough margin for other distribution channels, you can raise the price before you put the offering up on your site or sell through an e-tailer. Similarly, if you take preorders, you can raise the price when the preorder period has ended. In both of these cases, the perception is that you're rewarding early adopters with a discount and that the price has risen to the retail price once you've officially launched.

One thing you absolutely can't do is sell the product at a lower price on your company's site than your retail partners are charging for it in-store. Most retailers won't allow that, so if retail is something you're thinking of, it's important to build the retailer margin in right from the start. We'll touch on this, along with other concerns specific to particular distribution channels, later in the chapter.

In this section we present an overview of the three traditional approaches to pricing—cost-plus, market based, and value based—and discuss the pros and cons of each. We close the section with some further resources if you'd like to take a deeper dive.

COST-PLUS PRICING: A BOTTOM-UP APPROACH

Cost-plus pricing involves identifying all of the costs that go into producing a product, and then adding on the desired margin. We touched on some basic bottom-up product pricing calculations in Chapter 8; those assumed a simple "sell the device" business model and were designed to ensure that the costs of producing a perk were properly accounted for.

Bottom-up determination starts with accounting for your full cost of goods sold (COGS):

COGS = Materials + Labor + Overhead and Expenses

This is your *floor*. It's the lowest possible price you can charge and still break even.

As discussed in Chapter 8, this calculation is deceptively simple. Materials, labor, and overhead costs shift according to volume discounts and demand. Economies of scale might kick in as your number of units increases, but your initial assumptions must be conservative. While factory labor costs are rolled into the quote from your contract manufacturer, you'll have to account for additional labor required after your product leaves the factory. If you're handling shipping and fulfillment yourself, this includes the labor costs of the people who pack orders and take them to the post office. If there is a software component to your offering, this includes the cost of your developers. If you're a one-person team assembling and packing units on your kitchen table, don't forget to price in the cost of your own time.

Overhead and expenses might include rent on a warehouse and/or office space, bubble wrap for shipment boxes, gas, and postage. If you're establishing an online storefront, don't forget the cost of user fees for online platforms and payment-processing fees. Returns and defectives will impact your cost, and you should expect the defective rate to be high in the early days (perhaps as high as 10–15 percent). If you're manufacturing or shipping goods overseas, there are import/export fees and tariffs to consider.

In the bottom-up model, after you've calculated COGS, you add on margin:

Product Price = COGS + Profit Margin

Profit margin is partially a function of distribution strategy. If you're going to sell direct on your own site, you will capture a majority of the dollars over COGS. If you are going to sell through retailers, they will take a (potentially hefty) cut to satisfy their own margin requirements. You might view profit differently depending on whether you're planning to start a hardware business as your primary source of income, versus running it as a side project.

If you're a B2B company, sales cycles will have an impact on your profit margins. Procurement budgets, sales cycles, and decision makers vary across industries. Small and medium business owners (SMBs) are often able to decide to purchase from you immediately, without requiring additional approval, but they will likely do so at a much lower price point. Therefore, you'll have to make up for that lower price with increased volume in order to achieve profitability, so you'll need to have a plan for reaching many SMBs.

Conversely, large companies might have large budgets, but they often require several levels of managerial approval. This makes each contract a time-consuming process that involves a lot of hand-holding and sales skills. Simply identifying the decision makers in the company and getting a meeting on the calendar might take a lot of time. Therefore, pricing must take into account the time cost of the skilled labor required to get to a deal (this should be incorporated into COGS). If it's difficult to maximize quantity, it's necessary to maximize profit margin.

While this is an important exercise for thinking through your costs and establishing a floor, pricing experts will tell you that this is a terrible way to actually determine your price. This is because cost-plus pricing ignores the real world. It ignores market competition, and it ignores buyer psychology. The market simply might not bear the final price that you've arrived at using this method; this is often the case if you're competing in a crowded market against large incumbents who can take advantage of economies of scale. On the other hand, you don't want to risk leaving money on the table by pricing yourself too far *below* the competition if your differentiating features provide greater value to the customer.

MARKET-BASED PRICING: A TOP-DOWN APPROACH

Knowing your costs is critical, but pricing strategy must also take into account what your customers are willing to pay. This is a bit of a chicken-and-egg problem: since you haven't taken your product to market yet, you don't really know what they're truly willing to pay.

Identifying a price point that will enable you to run your company and turn a profit depends partly on the kind of volume you'll be doing. Sales volume is difficult to predict before you've gone to market, because it's a function of demand (gaining some insight into that demand is one of the upsides to a crowdfunding campaign). Unfortunately, until you go to market, it's often difficult to gauge the true demand level at given price points. At best, you have some indications from customer development studies, or from examining comparables already in the market. Be very conservative in your demand estimates in the beginning.

Market-based pricing involves looking at the prices that your competitors are charging, and selecting your price based on where you want to sit in the market. While cost-plus pricing is a *bottom-up* approach, *market-based* pricing is *top-down*: identify the competition already in the market, examine their prices, and select a price that is competitive with these other offerings. There might be multiple relevant axes of comparison—say, most features to least features, or most powerful to least powerful. The important thing is that you compare yourself to them in the same way that a customer contemplating a purchase would approach the decision. You must still know your costs, of course, because otherwise you could run out of money fairly quickly.

The primary concern with market-based pricing is that you may find yourself setting a price that doesn't leave room for a high enough gross margin. You simply might not be eligible for the type of volume discounts that a big company with an established supply chain can command. The widget that you pay a dollar to produce might be manufactured by a large competitor for 25 cents. If your competitor is charging $1.50 for that widget, you will have a hard time achieving profitability if you match the prevailing market price. Typically, the price for a product that will involve a retail distribution channel should be between two and four times COGS.

VALUE-BASED PRICING: SEGMENTATION MEETS DIFFERENTIATION

In his excellent pricing-strategy textbook, *Pricing and Revenue Optimization* (Stanford Business Books), Robert Phillips discusses the friction between the sales, marketing, and finance divisions of a company when it comes to establishing price. Finance organizations prefer the *bottom-up analysis*, which takes the company's costs into account. Sales teams like market-based pricing, because they can appeal to customers in terms of dollars saved. The third approach, *value-based* pricing, is the favorite of the marketing department. Value-based pricing looks at the price charged by the closest competitor (similar to top-down) and adjusts for differences between that competitive offering and the firm's own. That adjustment is based on the target customer's perception of the relative value proposition. A value-based model heavily weights the customer's needs; the idea is that if the product truly alleviates a significant customer pain point, she should be willing to pay a price proportional to the quality of the solution. Pricing to value is the preferred method of pricing experts.

To do a value-based analysis effectively, you'll be considering price in the context of how well your product solves a problem for a specific market segment. That means you need to know your differentiators—how you stack up against current market entrants—and whom you're selling to. When you're just starting out, you won't have historical customer data, so you'll need highly refined hypotheses about your customers and solid market data on competitor pricing across your space. Fortunately, you did most of this work when you identified your value proposition and brand positioning.

Regardless of whether you are pursuing a B2B or B2C business model, a value-based price will reflect the specific value you will provide to your customers. If your end customers are other businesses (B2B), investigate both what they're paying for their existing solutions to the problem (i.e., what your competitor is charging them) *and* what the problem actually costs them. Knowing the cost of the problem will help you identify a price ceiling; customers don't want to pay more for a product than they gain from having it. The price floor (discussed in "Cost-Plus Pricing: A Bottom-Up Approach" on page 223) and this price ceiling form a sort of bounding box around a feasible price range. This data might be hard to come by; most of the prices you'll see on your competitors' sites aren't the prices that they actually charge their customers, particularly if they're large. Many deals are individually negotiated.

If your target customers are individual consumers (B2C), knowing who they are is an important first step in quantifying the value you'll provide. Gender, income bracket, age, number of children, and location can help you formulate initial hypotheses about your target customer's budget for solving a particular type of problem. Psychographic profiles (sometimes called *behavioral profiles*) move beyond the "who" of the customer and try to get at the "why." As we discussed in Chapter 4, this type of segmentation groups people according to values, interests, and lifestyles (e.g. "the soccer mom," "the urban bachelor"). Your brand positioning is influenced by your target market; your price is a numerical representation of that positioning.

Let's say you're producing a wearable device for fitness. Are you targeting professional-caliber athletes, passionate hobbyists, or amateurs just looking to get off the couch? Those people might all be interested in a general class of product—say, sports watches—but are likely to want to spend very different amounts of money solving their pain point. The athlete might consider your feature-rich, highly precise sports watch a must-have and happily spend $300 on it. The amateur might find the pro-quality feature set appealing but the cost too high.

Once you've formulated a hypothesis based on market research, it's time to refine it, using customer research. Customer research typically entails conducting focus groups or interviews with individuals who fit various characteristics that you believe your customers (business or consumer) will have. The purpose of these conversations is to identify the specific pain points and needs of particular customer segments.

There is a right and a wrong way to approach pricing discussions during customer-development interviews. Don't ask questions like "How much would you pay?" or "Would you buy at $X? How about $Y?" Those questions are largely subjective and don't tie price to the value they will derive from your product. For a B2B product, questions should focus on unearthing the return on investment (ROI) of your product versus whatever the business is currently using. For example, if your robot can do the work of two human workers, understand what your potential customer is paying for his existing solution. Don't limit yourself solely to tangible benefits based on your feature set. Consider the experience factor as well. There are many cell phones on the market today, most of which perform the same general set of functions. Ease of use and a seamless, beautiful experience are factors that set the upper tier apart.

Even though ROI might not be as obvious for a B2C product, it's still possible to tie benefits to price by clearly conveying your value proposition. Ask your potential customers to articulate why they would purchase your product. Are you making something faster, safer, or better? Is your product something that conveys status?

Some entrepreneurs dislike direct-interaction customer research because they feel that customers don't really know what they want. They prefer to make a product in line with their own vision. If that's you, market research (rather than customer research) can help you determine a price. You can attempt to gauge what your customers will be willing to pay based on what your existing competitors are already charging.

Sonny Vu, founder of Misfit Wearables, took this approach. Personal sensors for fitness tracking have been around for a few years; the Fitbit, FuelBand, and Jawbone Up are three popular examples. Misfit studied competitors' products to identify both desired feature sets and feasible price points. Misfit's team read thousands of Amazon reviews for various products already in the market, meticulously noting praise for existing features, common complaints, and "I wish it did X" comments. They identified the competitors that their offering would most likely be compared to and price-matched so they would be on equal footing.

Selling It: Marketing 101

After you've identified a business model, found your market position, and determined an appropriate pricing strategy, it's time to formulate a marketing strategy to connect with customers. Many first-time entrepreneurs make the mistake of focusing a great deal of attention on the technical part of hardware entrepreneurship (building the device) and neglect sales and marketing until they have a pallet of devices waiting to be sold. That's a surefire way to lose a lot of money and waste a lot of time.

Earlier in the book, we dedicated a whole chapter (Chapter 4) to branding. That chapter preceded "Going to Market" by quite a ways because a founder should be thinking about brand identity during the formative stages of the company. In Chapter 4 we discussed the importance of knowing your differentiators and understanding the value your product can provide to a distinct segment of people. That learning from early on comes into play again here, as we discuss strategies for making customers aware that your product exists. The brand messaging that you worked to define is what your marketing activities are going to promote.

Marketing is the process by which a company communicates a product's value to potential customers. It's the art of making people aware that your product exists and making them want it. Any Marketing 101 textbook will include a mention of E. Jerome McCarthy (*http://en.wikipedia.org/wiki/E._Jerome_McCarthy*)'s "4 Ps of Marketing": product, price, place, and promotion. This framework was created by McCarthy in the 1960s. While there are new revisions with Es (*http://www.ogilvy.com/On-Our-Minds/Articles/the_4E_-are_in.aspx*) (experience, everyplace, exchange, evangelism), and occasionally Cs (*http://en.wikipedia.org/wiki/Marketing_mix*) (consumer, cost, communication, convenience), the classic mnemonic still summarizes the elements underlying a cohesive marketing strategy:

Product
What you're selling.

Price
What the offering will cost. As noted previously, this impacts product design, profit margin, distribution channel, and customer demographic.

Place
Where the seller makes the product available to the buyer (we'll discuss this later in the chapter, in "Distribution Channels and Related Marketing Strategies" on page 241).

Promotion
The set or series of activities designed to communicate the value of a brand or product to a target audience. Some of these activities include social media campaigns, the creation of content for a company blog, search engine optimization, public relations, and strategically designed product packaging.

Marketing activities are frequently divided into two categories: inbound or outbound. With *inbound* marketing, you "bring people in" through engaging content, podcasts, infographics, social media participation, community building, etc. Ideally, the content provides entertainment or educational value. The goal is the development of a longer-term, more two-way relationship with a customer. Inbound marketing is sometimes called *owned media* within the marketing industry, because the company "owns" (controls) the distribution channels. The type of marketing in which a customer has become a true fan and is proactively tweeting about, Instagramming, or otherwise sharing the product, is *earned media*.

By contrast, *outbound* marketing pushes a message out to potential customers through channels such as telemarketing, direct mail, or billboards. Advertising is a form of outbound marketing. Advertising typically involves paying a media entity—television, radio, a magazine or newspaper—to feature a message that's specifically tailored to persuade customers to buy the product. The goal of advertising is to buy awareness. In a visual medium, the message typically incorporates visual brand attributes such as the logo or typeface; this creates an association between the specific product and the brand. While content marketing (inbound) is produced with the intent to create value beyond selling a product, advertising (outbound) is about sales; money is exchanged for eyes. Outbound marketing is also called *paid media*.

There is no single correct way to run a marketing strategy. The mix of inbound and outbound activities that a hardware startup should pursue will vary from company to company. However, in the early days, money is probably tight. So we're going to focus on lean marketing strategies…not on how to connect with television stations to buy air time.

The market landscape research that you did preprototype helps set the stage for creating a marketing strategy as you get ready to sell. Broadly speaking, there are two types of markets: sellers' markets and buyers' markets. A *seller's market* is one in which the seller has a product that's unique, in high demand, or in short supply. A buyer of that product has to go to that seller. A seller's market is often called *product-focused*. In a product-focused market, the seller is the expert, the visionary. She can adjust the feature set and price with a high degree of flexibility because the product is not a commodity (something that is essentially equivalent across brands).

One example of this is the iPad: it was released, and suddenly millions of people realized they needed a $500 tablet. Many hardware startups with truly innovative new technologies find themselves in this situation. While it's nice to not have much competition, it's a challenge to interact with customers who don't yet realize that they need your widget. The business objective of marketing activities in a product-focused market is to increase market share and solidify dominance. By increasing market share, you increase revenues and reduce costs (assuming economies of scale affect your manufacturing process). This results in greater profitability. To increase market share, advertising should focus on the product itself.

The opposite of a seller's market is a *buyer's market*, in which there are many similar products that are equally desirable. In this type of crowded market, the buyer has the power. This is a *customer-focused* market. The customer is the expert, and the seller designs in response to the customer's needs in an effort to get ahead of the competition. The product is, first and foremost, a solution. If you identified a large number of competitors in your early research, this is the type of market you're in. As a result, marketing strategy and advertising efforts will focus on speaking to and connecting with a customer, appealing to his needs, and trying to instill loyalty. "That brand really understands me" is the goal. Since needs vary greatly from person to person, understanding customer segmentation is particularly important in a buyer's market. The seller can't be all things to all people; the goal is to do a great job of meeting the needs of whatever subset the seller targets. The business objective of marketing in a customer-focused market is to increase customer share (the percentage of the target demographic who choose your product) and solidify loyalty. As the customer-oriented company grows, it can eventually use this deep knowledge to produce and cross-sell other products that meet customer needs.

Now that we've covered some basics, let's work through a step-by-step framework for running a lean marketing campaign for an early-stage hardware startup.

STEP 1: DEFINE YOUR OBJECTIVE

In one to two sentences, identify the goal of your marketing campaign. Is it to raise awareness? Presell units? Develop your social media presence? You want something very clearly defined and not overly broad. In general, your objective will be related to a facet of either customer acquisition or customer retention.

Community is invaluable to user retention. In many sectors, it's easy for a competitor to knock off your design, or for the technology to become commoditized. Community, however, isn't so easy to knock off. If you have a strong group of loyal users and a solid brand identity, you have a more defensible position in the market.

Marketers often use the acronym SMART—Specific, Measurable, Actionable, Relevant, and Time-bounded—as a validation checklist to identify well-constructed objectives. Examples of such objectives might include:

- Increase product awareness among females 21–30 years old living in North America
- Close $200,000 in presales via company website in the three months following the end of our crowdfunding campaign
- Increase traffic to company website by 200 percent over the next month
- Grow social media base on Twitter, targeting 1,000 new followers and a 20 percent increase in mentions over the next 20 days

By contrast, simply saying "increase brand awareness" or "increase site traffic" is not SMART.

STEP 2: CHOOSE YOUR KPIS

Key performance indicators (KPIs) are measures used by decision makers to evaluate their progress toward achieving a goal...in this case, the objective you just defined. This is the nitty-gritty that underlies the *M* in SMART. Some KPIs, such as conversion rate or monthly unique visitors, are easily quantifiable. Others are a bit more qualitative or require some creativity. To measure an increase in customer satisfaction, for example, you might need to monitor sentiment in reviews. The important thing is to identify the KPIs that most accurately reflect your progress toward achieving your objective. For a crowdfunding campaign, you might be interested in average contribution, or number of backers per day. For a presale campaign, you might track site visits, revenue, or email list signups.

If you are new to KPIs, Shopify (*http://www.shopify.com/*), a site that helps small businesses build an online store, has a list of 32 popular KPIs for ecommerce (*http://bit.ly/32_ecommerce_kpis*). They are broken down into sales, marketing, and customer service–oriented metrics. Whichever you choose, make sure you're tracking the dollars you're spending relative to the success of your objective. Popular KPIs include cost per click (CPC), return on investment (ROI), customer lifetime value (LTV), cost per lead, traffic to lead (or lead to customer) ratio, and customer acquisition cost (CAC). Marketing costs money, and you don't want to spend it on campaigns that aren't working.

STEP 3: IDENTIFY YOUR AUDIENCE, THE "WHO"

Here, again, your early segmentation research comes in handy. If you're in a buyer's market, the "who" is obviously very important. But even in a seller's market, a young company with a limited budget will want to target a marketing campaign very carefully. If your objective requires reaching multiple segments, you'll want to provide relevant marketing materials for each one. That might mean buying ad space on a specific blog, or writing content to appeal to a particular demographic. "Campaign Page Marketing Materials" on page 172 in Chapter 8 covers this process for a crowdfunding raise; the general principles apply to most campaigns designed to raise awareness or generate sales. You want your message to be as tailored to each specific audience as possible.

Actual purchaser and user data is far more valuable than general demographic guesses. Marketers distinguish between *demographic segmentation* and *value-based segmentation*. Value-based segmentation is an analysis of the lifetime value of certain subsets of customers who have actually purchased your product. If you've already got a data set of customers who preordered on Kickstarter, think about ways to mine it for relevant insights that can help you achieve your current objective. In some subsets of hardware—particularly connected products that gather data—you can continuously refine your understanding of who your customers are and what they need.

STEP 4: SELECT YOUR MARKETING CHANNELS

Use what you know about your target audience and reach them in the places and with the methods that will resonate best. There are dozens of marketing channels. Some of the more common ones include email lists; ad space on relevant sites; Google AdWords; and social channels, including Facebook Ads, Promoted Tweets, Pinterest boards, YouTube videos, Instagram feeds, print media, or, for bigger budgets, television or radio. TVB (*http://bit.ly/tv_ad_cost_trends*), a trade association of groups associated with the US commercial broadcast television industry, tracks trends in the cost of TV advertising; in 2014, a 30-second spot on network TV during prime time cost $112,100 on average.

A full, in-depth discussion of each of these channels is outside of the scope of this book. However, since Facebook Ads, Google AdWords, and Twitter's advertising program are popular choices for lean marketing strategies, we'll touch on them briefly.

Facebook

Facebook's in-stream advertising is the mechanism of choice for many companies looking to promote their products. Software startups use it to suggest downloading their app, brands use it to drive users to their pages, and many hardware companies have made excellent use of Facebook Ads to drive traffic to crowdfunding campaigns and presale pages (see "Tile: A Case Study" on page 239 for a good example).

Facebook offers three possibilities for ad placement. The first is the right column of the user's page: the ads that appear there are targeted using simple keyword-driven advertising. The second location is *in stream*: the ads appear in the news feed, the same feed that displays posts by friends. The third possibility is the mobile news feed. According to Facebook's ad page, the Facebook app is installed on three out of four smartphones, giving it an extremely wide reach on mobile. A combination of better real estate and improved targeting options make the mobile and desktop in-stream ads more expensive than the right-column option.

Facebook has more than a billion users. Targeting involves selecting a series of criteria; the more selective you are, the better. Facebook doesn't want to spam its users or disappoint its business customers with poorly targeted ads, so it incentivizes businesses to create highly specific target profiles. For extreme specificity, Facebook's Custom Audiences product enables marketers to upload an email list to Facebook and target ads to those users. If you have a list of people who have entered an email address on your site but haven't ordered your product, this is a way to reach them. Facebook can also create *look-alike* audiences that match certain demographic traits of names in a list (or that match users who have Liked your Facebook page).

Businesses with budgets of any size can leverage Facebook Ads. Their own Success Stories page features entrepreneurs who have built communities for $5 a day. Well-targeted ads with highly specific target criteria are less expensive than broad pushes. Larger companies, including many popular ecommerce startups, spend well into six figures.

Google AdWords

AdWords is Google's advertising product. You buy keywords (which can be a single word or a phrase), and your ad is served in response to a query entered into a Google search field. For example, if you are building a fitness tracker, you might buy keywords such as "fitness tracking" or "run tracking" or "workout monitor." When a Google search-engine user looks for that phrase, your ad appears at the top or to the side of the results. It's also possible to make your ad appear within Google's Shopping or Maps feature or on partner sites.

AdWords is a *pay-per-click* advertising platform: a customer running an ad campaign is charged when a Google user clicks on the ad. Google allows you to set a maximum budget for an individual click (max CPC), and a daily budget across all clicks, and provides traffic estimates that take those budgets into account. If you aren't interested in driving traffic to a site and just want to increase brand awareness, you can also pay per *impression* (CPM; cost per thousand impressions) for ads on Google's Display Network.

Many books and blogs are devoted to mastering AdWords, so we'll keep our overview brief. Recognizing that it may seem daunting at first, Google provides resources to help customers get familiar with the platform. One of these tools is the Keyword Planner (*https:// adwords.google.com/KeywordPlanner*). This tool shows search volume for the keywords you have in mind. Using our example above, "fitness tracker" averaged 74,000 searches monthly between February 2014 and August 2014 within the US. The average for "Track my run" was 9,900 and for "workout monitor" was 390. The keyword planner is an excellent way to gauge demand in the early days (send people to a landing page even if you have nothing to sell), and to identify regional markets.

If hundreds of thousands of people are searching for a particular keyword, it's a competitive space. There's a finite amount of ad space on a Google.com search results page. Each time Google has an ad spot available, it runs an auction to identify which ad will get the spot. This is where *bidding* comes into play; although you are only charged if someone clicks on your ad, the price that you'll be charged is derived from what you bid to have your ad shown. You can either set manual bids or let AdWords optimize your bidding with its automatic system.

To work within a limited budget, narrow ad targeting makes the most sense. You can select exact-match or *phrase-match* modifiers to ensure that your targeting is as precise as possible. You don't necessarily want to choose the cheapest keywords. The more expensive ones are pricy for a reason; people who click those ads are more likely to become customers, so there's competition for them.

To create a campaign, you'll set a daily budget, select a target audience location (this can be extremely granular; you can draw an area on a map) and then choose relevant keywords. It's possible to include negative keywords, to avoid your ad showing up in an irrelevant context (for example, if a keyword is a homonym), and to serve ads designed for specific devices (e.g., ads that are optimized for mobile browsers). When someone clicks on your ad, you can track whether or not that person converts: a *conversion* can be defined as anything from viewing your site to purchasing a product.

Common metrics tracked by users of AdWords campaigns are ROI (in this case, income facilitated by the campaign relative to dollars spent on the campaign), click-through rate (CTR; the percentage of users who searched for your keywords who click on your ad), and cost per acquisition (CPA). A "good" CTR that indicates a well-targeted ad is between 2 and 30 percent.

Twitter

Twitter has an advertising program as well. In addition to helping you sell your product, Twitter is a powerful tool for gathering and managing your community. It's increasingly popular for handling customer service issues and facilitating two-way communication. Twitter has advertising products designed to grow your follower community and to increase brand reach through conversation (retweets, favorites, and @replies).

In addition to driving brand awareness, Twitter can help drive traffic to your site. There is a dedicated ad campaign type that enables a company to collect lead email addresses on Twitter itself (without sending people to an external site).

Twitter has several different audience-targeting strategies. You can target new customers based on keywords (either in their tweets, or that they search for), based on interests (as gauged by the accounts or lists they follow), or based on television programs that they engage with (yes, that's a distinct category; lots of Twitter users use the app on a *second screen* during favorite TV shows—for example, Game of Thrones reaches 823,000 people in the US). You can also target existing customers with the Tailored Audiences options.

Much as with the other platforms, you can focus regionally and by device, and you set overall campaign budgets, daily maximums, and individual bids.

STEP 5: FORMULATE YOUR MESSAGE

The goal here is to write a marketing message that will resonate with your target customer. Put yourself in your customer's position. What type of narrative will get that person's attention? Articulating her problem, and your solution, is a good place to start. Be cognizant of the tone and type of language you're using to communicate your story. Tell your customers why they should care, clearly and directly.

Marc Barros, founder of Contour Cameras and Moment (who appears in "Naming Contour and Moment: A Case Study" on page 69), had his fair share of messaging-strategy challenges with Contour. The team wanted to make a camera that enabled the average Joe to easily capture and share videos. "We struggled with whether to use a 'Be Like Joe' or 'Be Like Mike' strategy," he says. "We thought the end consumers' videos would spread virally; we were looking for that spiral effect on people's content to drive more camera sales." The team chose a messaging style that highlighted the everyman nature of its product.

Contour's competitor GoPro, however, employed a "Be Like Mike" strategy (a reference to the wildly popular Nike campaign featuring Michael Jordan). GoPro produced highly curated, highly polished content for its marketing campaigns. Its aspirational videos, shot by leading athletes and adventurers, sold the fantasy: use this product, and you'll have —and capture!—similar adventures and athletic feats. "The perception of what you could do with the product, the speed you could travel, the cliffs you could jump off, the richness of the video and the sound...this was stuff that the average person could never do," Marc says. "But it was the perception that sold it: 'Wow. My videos will look that good and I will look that amazing.'"

STEP 6: INCORPORATE A CALL TO ACTION

This is a specific action that you want your customer to take that will help you achieve your objective. It should be delivered as a clear instruction: "Call now!", "Stop by our new location!", "Buy today!" In the digital world, "Download the app!", "Sign up!", "Get Started!" are fairly common. Be specific and to the point—no vague "Submit" buttons! The call-to-action instructions for a single objective might vary by channel.

You want to make it as easy as possible for someone to take the desired action. For example, don't say, "Visit our crowdfunding campaign!" without including a clickable link. Also don't forget to optimize your site for mobile phone or tablet users. While some optimizations are fairly straightforward, hitting on the right language or method for getting a customer to perform the action may require a bit of A/B (or *split*) testing. For instance, if the objective is to get a customer to purchase a product from a preorder page, you might test a one-click checkout system that doesn't require registration alongside a more traditional order flow, and compare the resulting conversions. A product such as Optimizely (*https://www.optimizely.com/*) can help you test your calls to action.

STEP 7: SPECIFY A TIMELINE AND BUDGET
Most founders design early marketing campaigns around a set period of time (e.g., a two-month presale) or until a certain goal has been achieved (e.g., 1,500 units preordered). The SMART framework advocates for the former; the *T* of SMART is a set timeframe. If your goal is simply "Get 1000 new Twitter followers," you might plug away at that halfheartedly for months. With the focused objective "Get 1,000 new Twitter followers within a month," you've created a sense of urgency. Many founders find that deadlines help them maintain forward momentum. Either way, have a clearly defined end for a marketing campaign before it begins.

Similarly, have a clearly defined budget for a marketing campaign *before* it begins. With an unlimited budget, you would blanket the airwaves with high-production spots to drive people to your site. In reality, you probably have very little money to spend. Some channels might have greater impact but cost far more. Preemptively allocating your budget across marketing channels is a key part of the planning process.

STEP 8: REFINE YOUR CAMPAIGN
There's one area in which startups have an advantage over big companies in terms of marketing: startups have the freedom to iterate. A large company that hires an agency to run its marketing programs has very little opportunity to refine messaging and adjust budgets while a campaign is running. A startup, on the other hand, can do just that.

Before you start a marketing push, make sure you've got a plan in place to gather and analyze data around how it's doing. Your KPIs are tracking your progress toward your objective. Your analytics platform can track each channel's impact on those KPIs. If one channel is lagging, revise targeting or reallocate funds toward something that's performing better. Keep track of where traffic is coming from, how much it's costing, and where your message is spreading.

Google Analytics is free and incredibly powerful, and Google maintains the Google Analytics Academy (*http://bit.ly/analytics_academy*) with free online courses to help you become an expert. Customer intelligence platform KISSmetrics (*https://www.kissmetrics.com/*) has an excellent list (*http://bit.ly/analytics_resources_2014*) of more than 50 Google Analytics resources for all levels, broken down into topics like "Analytics for Conversion Rate Optimization," "Goal Tracking," and "Tips, Tricks, and Tools."

For a story of a team that pulled together all of the steps described in this section to create a successful marketing campaign for their product, see "Tile: A Case Study" on page 239.

Tile: A Case Study

Mike Farley and Nick Evans are the founders of Tile (*https://www.thetileapp.com*), a device that's designed to be attached to important objects so that you can recover them if they're lost. Tile had an extremely successful Selfstarter campaign, selling 200,000 units, and has since gone on to sell over 500,000 units in total.

While discussing their approach to marketing, and "crossing the chasm" into mainstream audiences, one word came up repeatedly: *simple*. "We wanted a very simple, elegant offering that was really easy to understand. That was the goal," Nick says. This emphasis on simplicity anchored both product design and product marketing. To get started with Tile, you simply attach it to the object you care about, such as your laptop or your keys. Tile then acts as a beacon if the object is lost; the tracking signal spans a 100-foot radius, and any phone running the Tile app can help locate the missing object. The app's tracking signal gets stronger the closer you are to the object.

The team began their marketing push by using Selfstarter to host a presale campaign on Tile's own site. The assets developed for the campaign, including the video, remain the primary marketing materials on the site today. "We focused on making the video about the solution to the problem of losing things. We stated what Tile was early on and just got straight to the point," Mike says. The team didn't discuss the technical specifications in the video. Clarity was key. "We didn't want to muddy the waters with things that customers weren't really going to care about," Mike says. "Most people don't care if it uses WiFi, Bluetooth, or GPS. They care whether or not it solves their problem." They decided to make the messaging as simple as the product.

At the time of the launch, the team was still small, with only four full-timers. Cofounders Nick and Mike were the only two who worked on marketing. "We could only do so much, and we knew that it was extremely important to get out into the magazines," Mike says. "We did our research on previous crowdfunding campaigns and one of the things that really added credibility was when people got coverage in important publications." The team wanted to build the kind of credibility that would carry them through the Selfstarter campaign and beyond. It was important to build a reputation and brand that would resonate with a broad customer base, since their product had a large target market: people who lose things.

Besides credibility, getting good press hits was a vital part of getting the word out. The Tile team made the decision to hire professional PR team VSC (now Wareness.io), which has gone on to help launch successful projects such as Coin and Ringly. "We got creative with limited funds early on," Mike says. "We had $200,000 from the Tandem accelerator program to finish building the product and get to launch. We used every single penny of that $200,000." The VSC team would occasionally chime in with some marketing advice, but for the most part they just handled PR.

Digital marketing was the most effective way to get the best results with a limited budget. The team created a basic marketing plan. They set some basic KPIs and had a spreadsheet to make sure they were hitting their numbers. "Though, it was a pretty primitive spreadsheet," Nick says.

The team relied heavily on Facebook advertising to drive people to their site. They put ads out on Facebook, emphasizing the simple elegance of their product. People began to share Tile with their friends. "People wanted to share it because of the simplicity, and because we were solving such a universal problem: losing stuff. Everyone knows what that feels like. That combination allowed us to cross that chasm early on," Nick says. The team had a contract manufacturer lined up to produce 20,000 units; over 200,000 were ordered during the initial campaign.

To continue their momentum, the team has since stepped up their Google AdWords strategy as well. As of October 2014, over half a million units have sold, and the team is preparing to ship version 2. The team credits their unwavering focus on the product and value proposition for their advertising success. "You can get better and better at ads, but there's no big secret to how we did our early digital ads," Mike says. "We just got to the point quickly. This is the product, this is the value proposition. Everyone is so easily distracted now, you have to get to the point real fast." Getting users to your site is only half the battle; making it immediately clear what you do, what problem you're solving, will help you make the sale.

It's important to note that the reason for Tile's success is that there's a great product underlying the ad campaigns. "This is so foundational, but you really have to start with a product that people want," Mike says. "You have to be super honest with yourself about that. We were able to conceive of a product and build it, and real customer need drove the campaign." Simplicity can help people to understand your offering. Digital marketing and effective PR can help you cross the chasm. But ultimately, even the best marketing strategies and ad campaigns won't be able to sell a product that no one wants.

Distribution Channels and Related Marketing Strategies

The last of the "4 Ps of Marketing" to discuss is *place*. Distribution channels are the avenues through which a seller makes it possible for customers to purchase the product. It's important that your price, distribution strategy, packaging, and messaging are cohesive. All four Ps should fit with a specific target customer identity.

When we talk about place, we're asking where that target customer typically shops. Does he frequent upscale department stores? Trendy boutiques? Online electronics sites?

At the broadest level of evaluating distribution, you can make a sale to a customer either online (ecommerce) or offline (in a brick-and-mortar store). Ecommerce can happen using a direct-sales strategy via the company's own website, or through a site that specializes in aggregating the wares of other brands such as Etsy or Grand St. Brick-and-mortar retail stores run the gamut from smaller mom-and-pop specialty stores, to big-box retailers with a national presence. For B2B products, distribution often happens via direct sales.

ONLINE DIRECT SALES

The most common way hardware startups begin selling their products is through direct sales on their websites. Ecommerce sites are inexpensive and easy to get off the ground; you have full control of inventory, pricing, and sales strategy; and there's no one to split margin with…it's all yours!

An increasing number of out-of-the-box software solutions will get you up and running almost instantly and at relatively low cost. An ideal solution should enable you to manage inventory, accept payments, track orders and shipments, and easily put items on sale. Shopify (*http://www.shopify.com/*), Volusion (*http://www.volusion.com/*), and Magento (*http://magento.com/*) are popular choices for quickly creating a basic storefront (though there are many other options). You'll also want a high-quality analytics package to gather data about user behavior on the site and track sales KPIs. Make sure you're able to tie sales to customer service inquiries and returns processing (you need to know what your real return and defect rates are before you move into other channels). Should you choose to build your site framework from scratch, a number of plug-and-play shopping carts and PCI-compliant payment providers can handle the more complex parts of the checkout experience.

The primary challenge of direct sales is that you have to do extensive marketing to generate demand. The web is a noisy, cluttered place, and depending on what you're selling, you might be unlikely to rank highly in search results initially. This is one of the reasons why it's so important to build community as early as possible. If you're going to do a crowdfunding raise first, you want to be sure that those first customers and earliest evangelists know how to find your company after you leave the crowdfunding platform. If you didn't do a crowdfunding raise, you'll want to start building an engaged customer base immediately.

Many startups begin with direct sales before they've raised venture funding. At a bootstrapped company, this often means the marketing budget is extremely tight. For online direct sales, growth-hacking techniques employed by ecommerce startups make effective marketing strategies; there are many blogs devoted to growth hacking for ecommerce. Postmanufacturing, the sales challenges of a new hardware business look quite similar to the challenges of a small clothing brand.

When a customer places an order or preorder, collect an email address. Create a periodic newsletter to keep previous visitors up-to-date on new offerings, and design a *drip campaign*: a series of emails that happen in a set order, on a set schedule. Engage the community in some way, perhaps a survey about what color or feature you should incorporate next. Women's apparel startup ModCloth (*http://www.modcloth.com/*) has a wildly popular "Be the Buyer" feature that solicits feedback on potential new inventory and presells it.

Turn your customers into evangelists. Another apparel startup, Betabrand (*http://www.betabrand.com/*), uses its customers as models with a feature called Model Citizen: customers take funny photos of themselves wearing a Betabrand garment and upload them to the site. The shots are featured on the relevant item's product page and are easily shareable with friends. These photo testimonials drive engagement, facilitate new customer acquisition, and are a well-integrated form of social proof.

Use sites such as Facebook and Pinterest to create an accessible brand identity and reach new markets outside of the tech community. Strategic press coverage in publications (online or paper) that address your target market segments is important as well. If you have the budget, consider hiring a professional PR team; if not, develop relationships with reporters. If your business is solidly B2B, consider writing case studies highlighting specific ways that your product has solved a problem for users in various verticals. Ask for testimonials from prominent customers, and highlight them on your site. Create searchable content and thought leadership related to your problem space.

Metrics That Matter

If you're looking to build a small or lifestyle business, direct sales might be where you choose to stop. However, if you're planning on using sales from your storefront to demonstrate demand to bigger retailers or investors, make sure you're tracking the metrics that they care about. Pageviews matter: how many eyes are on your site? Track conversions: what percentage of visitors go on to buy? Measure click-through rates from advertising channels, as your cost to acquire a customer fundamentally impacts your bottom line.

Many ecommerce platforms include analytics dashboards so you can see how your customers find you, monitor trends, project revenue and sales growth, and optimize accordingly. Kissmetrics has a particularly strong focus on ecommerce customer intelligence, and it regularly publishes blog posts and case studies detailing best practices for growth (*http://bit.ly/growth_best_practices*). A/B testing of product messaging, layouts, and cross-promotion can also help you optimize.

Offline metrics also matter. One important metric is the Net Promoter Score (NPS), a gauge of customer loyalty to a company or satisfaction with a particular product. It's measured on a sale of -100 to +100. The NPS is calculated using a customer survey with one simple question: "How likely are you to recommend this product (or company) to a friend, on a scale of 1–10?" The idea is that loyal customers, in addition to providing you with repeat business, will be likely to recommend your product to their friends. A promoter typically responds with a 9 or 10; a more passive customer, a 7 or 8. Anything less than a 7 is considered a *detractor*. To get from the numerical result to the NPS score on the -100 to +100 scale, you subtract the percentage of detractors from the percentage of promoters.

Satmetrix, the developer of the methodology, tracks industry benchmarks. While they vary by sector, a +50 is considered excellent in most cases. In 2014, the category leader for laptop computers was Apple, with an NPS of +72. This isn't an across-the-board rating; it's specific to the product category. In the tablet space, Apple's NPS for the iPad was +66; in smartphones, it was +67. NPS is used to track general perceptions of an entire industry (e.g., technology, or banking), specific companies within the industry, and specific products within the company.

Proactive community management can do wonders for your NPS, particularly in the early days. Software startup marketing folks often talk about *viral loops*. This is an extension of the principle underlying the NPS; it's the idea that you can encourage your first 20 users to each refer 20 new users, and then those new users each refer 20 more, and so on. Virality is most likely to happen if the user experience of the product is in some way improved by having more of one's friends also on it. In hardware, this is a bit more difficult to achieve, but it can still be done. Eric Klein, a partner at hardware accelerator Lemnos Labs, was previously the Senior Director of Product Marketing at startup Dash Navigation. At Dash, he began to use NPS while the product was still in the DVT stage. "We would do an NPS analysis each time we released a new DVT," he says. "We created cohorts, and resampled NPS as our product evolved, even before we hit PVT." This gave the team a sense of how likely consumers were to tell their friends about Dash long before it hit the shelves.

Though this discussion of metrics falls within the "Online Direct Sales" section of the Distribution portion of this chapter, key metrics should be tracked across any and all channels that you leverage. It's important to build granularity in from the beginning so that you really understand what's happening as your company grows. Eric explains, "NPS is supposed to be blind to channel, but with different channels you will have different customer experiences. You need to differentiate your score. If you find that direct sales is a +72 but mass market retail is only a +55, that tells you that customer expectation is being set improperly between the two channels." One way to track this is to create special identifiers that act as markers, such as when your customers register their product. You want to know which batch was sold direct, which went to a retailer, etc.

ONLINE SPECIALTY RETAILERS AND RETAIL AGGREGATOR PLATFORMS

Etailers can be a great way to leverage the audience of an existing platform to call attention to your own brand and sell in a relatively low-touch way, particularly consumer products. In this section we're not talking about the ecommerce sites of behemoths such as Target.com or Walmart.com; we mean online-only specialty shops—niche sites that cater to hardware geeks, early adopters, or electronics lovers.

Some of these sites, such as Newegg, Adafruit, and SparkFun, act like retailers, purchasing inventory at wholesale prices and reselling it. Others, such as Tindie or Grand St., act as *aggregator* marketplaces. Tindie, a popular site for maker products, allows anyone to hop on and list items in a model reminiscent of eBay Stores. Grand St., recently acquired by Etsy, markets a new product every few days and includes Featured Preorder and Featured Beta channels to help very young companies find their audience (see founder Amanda Peyton's discussion of leveraging an etailer's audience in "Grand St.: A Case Study" on page 245). Finally, flash-sale sites such as Gilt or Zulily are another potential ecommerce option to consider; target one that serves your customer demographic.

Grand St.: A Case Study

Establishing distribution channels is a critical part of turning a hardware project into a bona fide company. Even if a startup has used a crowdfunding raise to generate buzz and sell a first batch, it needs a clearly defined strategy for continued sales. In this case study, Grand St. founder Amanda Peyton discusses distribution—and the data that matters—for young companies.

"Grand St. began when we realized that there was no great place to discover and purchase small-batch or independently made electronics," Amanda says. Most small hardware startups aren't able to produce the hundreds of thousands of units that are required for distribution by major national retail chains. They're still concerned about runs of a few thousand and nailing down product-market fit. At the same time, they have to come up with a strategy for sales. Typically, that involves an ecommerce platform and direct-to-consumer sales. In those models, discovery can be a challenge.

Grand St.'s vision was to create an online hub for selling the type of products that resonated with an early-adopter tech-loving audience. "We wanted to create a compelling, community-driven marketplace where people could sell the units they were creating, something really friendly to new hardware companies," Amanda says. Grand St. began by featuring high-quality, unique products in limited-time sales and gradually expanded into an ecommerce hub where founders could set up their own shops. The site currently still sells items on consignment via promoted sales, but it increasingly encourages founders to leverage the community and audience while shipping their own products.

Grand St.'s aim was to be an ideal first external sales channel for hardware founders. The site helps companies gain exposure to new audiences through Grand St.'s own marketing. Storytelling is extremely important. The team writes marketing copy for the items they feature, explaining how the product fits into a customer's life (something, they've discovered, that many founders neglect). They promote the products they feature on social media channels and news sites. They also do in-person marketing at events and pop-up shops.

Amanda emphasizes the value of data in refining marketing. The team extensively analyzes data from its customers to provide feedback to founders on everything from their marketing materials to their price points. "We can tell founders about the audiences that their product resonates with—what sites they are coming in from, their gender breakdown, what other categories of products they might consider and at which price points. Sometimes it's not who they thought," Amanda says. This data can help a startup to more efficiently market its current product. It can also help drive development of the next version.

In thinking about distribution, and whether to sell your product exclusively on your own site or on a platform like Grand St., margins are obviously an important consideration. However, it's also important to remember that driving traffic to your storefront requires a considerable amount of time and talent. Aggregated ecommerce distribution channels like Grand St. can be a valuable way to grow your audience in the early stages of building your business.

Etailers aren't free. Charges will come in the form of either monthly or per-listing fees, a percentage of sales revenue, or some other formula. Be sure that your price is set at a point where you can still turn a profit on sales from each site. You'll also have less control over the purchasing experience, since someone else is handling site design, marketing, and customer satisfaction. Fulfillment might be a point of concern. Some etailers hold inventory in their own warehouses, and others require you to ship directly to the customer after they've sold on your behalf.

When a retailer takes on inventory, you are in what is known as a *sell-through* relationship. Let's define *sell-in* and *sell-through*. In a *sell-in* transaction, the retailer buys the goods from you at a discount and then sells them to the consumer. You send an invoice to the retailer when you ship it the goods. The retailer reserves the right to return unsold inventory to you in exchange for a refund or credit. The *sell-through* happens when the consumer pays the retailer; you have *sold through* the retailer to the end consumer.

Monitoring both sell-in and sell-through numbers is important. Sell-in data reflects the number of units you've put into the retail channel, whereas sell-through is how many have actually been purchased. Ad campaigns and placement in store circulars or email campaigns are designed to increase sell-through. In this type of relationship, if units don't sell, they are likely to wind up back in your hands.

Amazon.com is one of the most popular online retail destinations. It acts as both a marketplace for individual storefronts and as a retailer. Selling in the Amazon Marketplace costs $39.99 per month for most professional-tier stores that handle physical goods (a *professional* seller is someone who sells more than 40 items a month). There's also a referral fee that ranges from 6 to 25 percent of the price of the product, depending on the category (*http://bit.ly/amazon_selling_fees*). If you're selling downloadable software (or other media), there's an additional Closing Fee of $1.35. For an additional set of fees, Amazon offers its marketplace sellers access to the Fulfilled By Amazon service, which handles pick, pack, ship, and return services. We'll discuss this more in "Warehousing and Fulfillment" on page 261.

Many online marketplaces simply take a percentage of revenue sold on their platforms, or charge a flat monthly fee. If the online vendor is acting as a retailer, however, the fee structure is different. If you enter into an agreement with an Amazon buyer to have Amazon act directly as the retailer through its Vendor program, the fee structure changes. Its gross margin ranges from 25 to 50 percent, depending on the sector. They also offer cooperative marketing programs such as promotional pages, branded sites, coupons, and more (the cost for these offerings is similar to the big-box retailer market development fund costs that we'll discuss in "Big-Box Retail" on page 249). Items that are sold by Amazon.com directly carry the designation "Ships from and sold by Amazon."

Let's briefly touch on markup versus margin: your *markup* is the difference between what it cost you to produce a widget and what the retailer paid you for it (the wholesale price). If your product costs $10 to produce, and Amazon buys it for $25, the markup is $15, or 150 percent ($15/$10). Your *gross margin* is the markup divided by price the retailer paid for it—in this case, $15/$25, or 60 percent.

Amazon cares about its gross margins as well. If it paid $25 wholesale and sells the item for $40 retail, its markup is $15 and gross margin is 37.5 percent ($15/$40). Gross margin targets vary across different types of retailers. As mentioned in "Pricing" on page 220, it's important to know the requirements for the distribution strategies you're considering. If you use a *distributor* (a company that acts as a middleman, buying product from you and selling it to a retailer) to get your device into a particular channel, there will be an extra level of margin.

If you choose to go with an online marketplace or etailer *before* hanging out your shingle, verify that you'll have access to data about your customers, along with a way to communicate with them. You will likely want to stand on your own at some point, so don't grow overreliant on the sales and marketing engine owned by the retail platform. It's important to have a timeline in place for building out your own sales and marketing teams.

It's important to do your homework when making your first forays into a retail relationship. You need to understand how the buyer thinks, to know how you'll finance inventory to meet retailer demand, and to properly budget for market development funds (MDF).

SMALL RETAILERS AND SPECIALTY SHOPS

This distribution channel is a great bridge from online-only into the world of brick-and-mortar sales. "There are two things in between etail and mass market big-box," Eric Klein says. "The first is regional sales, which is similar to mass market but on a smaller scale. The second is specialty shops." Dash Navigation went with a regional electronics store, Fry's, to try out the mass market experience without having to produce tens of thousands of units. "You're still one of many SKUs in the store, and there are higher volume requirements, but there are also fewer stores," Eric explains. "You don't have to put as much volume into the channel [as you would with big-box], but things like market development funds and endcaps are handled the same way."

Small and specialty retailers often have a clearly defined customer base, so you can continue to sell to your niche while learning about the retail process. There's less concern about getting lost on the shelves of a specialty retailer versus a larger store. It's a great opportunity to explore *unaided sales*: selling without a company representative or website marketing copy immediately available to clarify the offering for a customer.

It's important to start your brick-and-mortar experience with partners that will give you feedback about customer experience of your brand and provide you with insights that can help you improve your product. "One of the scariest parts of being a hardware entrepreneur is the first time you put a lot of inventory into a channel," Eric says. If you sell into big-box and your packaging is off, marketing is poor, or the product doesn't resonate, the retailer will return the unsold units to you and not want to stock your product again. In smaller stores, the buyer and manager may be the same person. This individual can tell you if customers understand what your product does, share questions that customers are asking, and provide feedback about how effective your marketing or packaging is.

If you don't want to fully commit to brick-and-mortar just yet, a pop-up store might be a good way to give it a try. Particularly around the holidays, unused retail space can be rented out for much less than the cost of signing a lease on a commercial location. Storefront.com, which calls itself "the Airbnb of retail," is one place to locate a temporary or shared space.

BIG-BOX RETAIL

Big-box retailers with a national presence often seem like the holy grail of distribution to early-startup founders. They get a lot of foot traffic in their physical locations, and their own brand recognition means that their websites get a lot of direct traffic. Large retailers can generate exposure and theoretically drive a lot of sales, both online and offline. However, there are a number of pitfalls to avoid. In order to be on the shelf of a major retailer, your product has to be perfect. Retail customers aren't Kickstarter backers; they expect a refined experience and will be annoyed, disappointed, or angry if they have to waste time returning your product. Retailers will monitor returns. Not only will you be on the hook for replacement or repair costs, but they will also cease to carry your product if the percentage of returns climbs too high.

Jason Lemelson of Slam Brands (see "Slam Brands: A Case Study" on page 250) articulates just how different the big-box sales experience is:

> There is no relationship between selling online versus selling to a small retailer versus selling to a national chain. The process is different and distinct and there isn't a lot of crossover in the skill sets required in those different channels. Managing a handful of SKUs on a specialty site doesn't help you get any closer to what's required to service Walmart operationally. There's a sophistication level and scope of operations that's on a different scale. There are product testing requirements and certification specifications unique to each retailer, manufacturing and logistics compliance issues. I am unaware of anybody who is selling to a national chain that doesn't have a significant presence in China.

Slam Brands: A Case Study

Getting your product onto the shelves of big-box stores can be incredibly time-consuming, and the path in is often opaque. Jason Lemelson, founder of Slam Brands (a gaming gear and furniture brand), has spent over a decade navigating the pitfalls. He's placed Slam Brands products on the shelves of Walmart, Target, Best Buy, Costco, and more, reaching international distribution in more than 15,000 locations and doing millions of units in transaction volume. Here, he talks about how important it is to know what you're getting into when you target the big-box stores.

Slam Brands was founded in 2000 and initially focused on manufacturing, designing, and distributing ready-to-assemble furniture. Over the years, the company expanded into gaming accessories and wood furniture, which led Jason to form relationships with many different types of retailers. At first, he found the process extraordinarily difficult. Many buyers weren't interested in talking to startups or small companies, believing either that the product wouldn't be polished or that that the small company would have trouble meeting retailer demand. However, he persisted. "Getting national chain distribution is an essential aspect of scaling a business," he says, because "so much money is made in the top 10 retailers."

Small companies often use *rep groups* (sales representative agencies) to initiate relationships with national chains. The right rep group can help a founder get an audience with a buyer that might otherwise be difficult to approach. They work on commission, and at the national chain level, the fee is typically in the 2–10 percent range, depending on the product category. To keep the representative incentivized, compensation is often tied to sales volumes.

Rep groups can also help startups understand the buying cycle for a particular partner. The cycles are extremely complex and difficult to understand from the outside. Each of the major chains has a specific set of processes for merchandising, and they often move slowly. Jason explains, "Each buying cycle is unique, so as we scaled up, we had to evolve our operation to a point where we had specified account managers who tracked the buying cycles of each of the respective national chains."

Slam Brands had an internal industrial design team and engineering team, and they divided design cycles according to the targeted store. Many items were designed around in-store date targets as far as 18 months into the future. "Lots of new companies go out and build a product and then start to think about how they'll sell it at the end of the design cycle. That's backwards," Jason says. "Companies who are selling to these retailers are designing SKUs specifically for them." SKUs built for Target were designed with the demographic, average income, and appropriate price point for a Target customer in mind. They were completely separate and distinct from those being developed for Costco and Walmart, because each company wanted unique merchandise.

As a result, all discussion about price points and design happened within the context of both the wholesale and retail customer. The Slam Brands team constantly asked themselves, "Why would Walmart carry this?" They conducted strategic merchandising analysis in the categories they cared about and researched competitor price points. Once they had a sense of where a product would fit and what price was reasonable, they would subtract out the margin for the retailer and back out what they had available for their bill of materials (BOM). Packaging for big-box retailers was also an important part of the product design process, as Jason notes: "Designing a box that looks good on the shelf, where a mom with two kids in tow will be able to look at it, know what it does, and decide to buy it—that's a significant challenge." Every inch of a retail shelf is carefully planned out.

There is also the challenge of balancing volume and exposure. Large retailers place orders in the hundreds of thousands to millions of units. Slam Brands found that smaller orders across multiple ecommerce sites took as much time to manage, but brought in much less revenue. Jason remembers, "We liked the idea of using ecommerce to pioneer a new product, but once you've structured your business to have what it takes to land Walmart and Best Buy, it's difficult to focus on the needs of the little guys."

> Deciding to target large retailers is a big commitment. It potentially means not having the bandwidth to service smaller partners.

Before seriously considering retail, it's critical to understand the costs (see "Nest on Distribution Channels: A Case Study" on page 252). The capital requirements for retail are huge, cash flows can be complex, and the combination of those factors can kill a startup. There are three primary cost issues to be aware of:

- Retail gross margins are typically in the range of 30–50 percent and can vary by department within a single retailer. Your price has to account for both this margin and your own operating needs if you are going to turn a profit.

- Retailers will require you to deliver a large amount of inventory up front, often in the hundreds of thousands of units. You have to pay for the manufacturing costs and transportation costs to the retailer's distribution center, for somewhere in the neighborhood of tens of thousands to millions of units.

- Retailers' payment terms are, at best, typically what's known as *Net 30*. You'll send an invoice when you send them the inventory, and they'll pay within 30 days of receiving the invoice. Since you're getting paid after delivery, you have to front the cost of production. This ties up capital and can make scaling extremely difficult.

Nest on Distribution Channels: A Case Study

In "Nest Branding: A Case Study" on page 75, Nest's Matt Rogers discussed the company's philosophy around brand building: think about it early, and make it a priority. The founders also set their sights on a retail distribution strategy right from the start. They met with representatives from potential retail partners long before launching their first product, and when setting price, they took retailer margins into account.

The prospect of meeting retailer demand can be daunting for a new startup, particularly one with a hot new product that's seeing a lot of demand. Retailers are making a guess as to how many units they can sell, and they're often wrong. The question is, which direction are they wrong in? A retailer might believe it can move a very high number of units a month, but it bears none of the risk. If they don't sell, it will return the product to you. "If you deliver a million units a month, it's on your backs if it doesn't sell," Matt warns. On the flip side, the retailer may underestimate demand, leaving your product backordered until you can produce another run.

Since the company's working capital came from its venture round, Nest built conservatively in the early days, and it allocated units conservatively across its retail partners. "That meant there was scarcity for the first two months, but that's okay," Matt says. "It's better to be conservative and not spend all your capital than be stuck with no money to run your company." When the team first launched with Best Buy, they started in three retail locations before deciding to scale up. Nest and Best Buy chose the locations together: the Bay Area, Chicago, and Austin. They used analytics to compare city demographics to their target market.

When they launched the Nest Protect, their second product, they approached their retail partners a month before the launch. They'd kept their plans a secret even from the retailers carrying the Nest Thermostat. Only when it came time to start conversations about shelf space did the Nest team reach out. They not only told the retailers that they should carry the Protect, but they also suggested where it should go on the shelves. "We did a lot of planning for them ahead of time so it was an easy decision to make," Matt says. "It all comes down to negotiation and placement. When they're doing their store designs, be involved in that process." If the retailer believes the product will sell well, it will want it to be prominently placed. Getting the endcap (the product display at the end of the aisle) can provide a good deal of lift for a product and is worth negotiating for.

Getting onto the shelf is only the beginning. The Nest team is in constant communication with retailers about placement, promotions, marketing programs, and store performance. They support their retail distribution with market development funds (MDF). The team's preference is to pay for performance, or to pay for actual marketing. Matt says, "I really don't like flat MDF of the 'Pay us 3% extra profit on every unit we sell' variety." The team instead ties MDFs to specific programs.

Hiring someone who has had experience selling into retail is critical for getting the best possible vendor agreement in place. "When it comes to retail, don't let them drive. You have to control your own destiny," Matt says. "Dealing with retailers is a specialty in itself. You'd never hire any old plain vanilla software engineer to build hardware; you hire the guys who know how to build hardware." Expertise in retail negotiations is an important skill; an expert will know what's negotiable and what isn't.

Ultimately, when considering an arrangement with a retailer, it's important to fully understand everything that the deal will entail. "The onus is always on the company to do the right thing, to plan for their own financial success," Matt says. If you're going to sell to big-box stores, roll out gradually, manage your cash carefully, and hire the right team.

Even if you're capable of navigating the financial hurdles, getting onto the shelf is still a challenge. If your startup attracts a substantial amount of buzz, retail buyers may reach out to you. Otherwise, you are reaching out to them, often via cold email or cold calls. Trade shows such as the Consumer Electronics Show (CES) might provide an opportunity to connect. There are many vertical-specific trade shows that can help you reach buyers for your target audience. However, much like meeting with investors, relationships with buyers are built over time. Retail buyers often aren't especially interested in forming relationships with companies that only have one SKU to sell unless the product is unique. It's particularly difficult to sell a product in a sector already dominated by big brands with existing shelf space.

Since buyers get hundreds of inquiries a day for new product placements, rep groups are often the easiest way in to a retail store (see "Rep Group Tips and Tricks" on page 254). A rep group can help alleviate the friction that young hardware startups typically encounter when dealing with buyers directly. The startup often has no established sales organization that can take the time to meet repeatedly with buyers to help them understand the product; a rep group acts as that sales organization. Rep groups have friendly relationships with buyers, and have a strong sense of what a retailer is looking for. As multiline salespeople, reps know about a lot of products across the range of categories of products they sell.

Rep Group Tips and Tricks

Chris Mason, founder of Intelligent Products Marketing (IPM) (*http://www.ipmrep.com/services*), and Steven Levine, founder of Next Level Sales and Marketing (*http://nextlevelsales.com/*), share their expertise.

What does a rep group do?

Rep groups are independent sales professionals who represent your product and help bring it to the retail market. Reps may additionally offer value-added services such as providing category-specific market research and sales trend data, setting up displays and demonstrations, and training retail employees in how a particular product should be used. Rep groups may work alongside in-house sales teams, or they may be used in lieu of such teams by a small business or young company. Because of their extensive knowledge of the quirks and terms of individual retailers, companies often hire a rep group to assist with setting a price and designing packaging. Many retailers have specific packaging requirements.

Why should I use a rep group?

"First and foremost, speed to market," Chris Mason says. Reps know how to do business with major retailers. They have relationships with buyers at stores of all sizes, both online and brick-and-mortar. IPM, for example, focuses on selling to West Coast and online retail accounts, including Amazon.com, Apple, Costco, and Walmart.com.

Steven Levine emphasizes the market expertise that a multiline sales professional can offer. Next Level Sales helps electronics and housewares vendors sell into many retailers, including Newegg.com, Fry's Electronics, and HomeGoods. "Reps know a lot of products across a bandwidth of categories within the channel they sell," Steven says, "They are the eyes and ears of their sales territory." Reps have a deep knowledge of retail opportunities in their specialty area. They typically form strong networks with other reps across the country and the world, and can leverage those relationships to increase the breadth of distribution or open up specialty channels (such as military bases).

It's important to note that reps won't represent just any product. They pride themselves on showing their buyers high-quality and relevant goods, and they don't want to risk their own good name bringing in widgets that are poorly conceived or a bad fit. Reps also stay within their domains of expertise and are likely to turn down products outside of those areas. It's simply too challenging to maintain high degrees of market awareness and strong relationships across all channels. Specialization is key.

When should I contact a rep group?

Reps like to begin working with a vendor (in this case, a hardware startup!) as they're getting ready to go to market, four to six months out at the longest. Occasionally, a rep group will start helping in the prototype stage, but that's rare.

Before approaching a rep group, be sure you have given a good deal of thought to your ideal retail partner, the size of order you can fill, and the sell-through potential of your product. Knowing your budget for promotions and advertising is key. Having strategies in mind for driving customers into the store to buy the product is also critical.

What is a typical process for working with a rep group?

A vendor will typically approach a rep group with a looks-like and works-like prototype, or a series of artist renderings. The vendor will mention a few retailers that the company is interested in reaching out to. The rep will discuss the retailer's margin and product-delivery requirements, payment terms, and how well the vendor would fit.

After the vendor signs a contract with the rep group, the rep reaches out to retail buyers. She will often begin by sending a glimpse of the product to a buyer, setting up a teaser that will resonate well. Typically, the teaser includes the product concept and a company summary. Upon securing a meeting with the buyer, the rep will show the full deck, demonstrate the product functionality or show samples, and discuss the vendor's go-to-market strategy. Depending on the stage of the product, the rep will stay in touch with the retailer, showing the evolution of the product and packaging, or keeping the retailer updated on the ready-by date.

If the retailer decides to carry the product, the rep negotiates the vendor contract. Most of the time, the retailer has a set vendor agreement and there is little room to negotiate, but occasionally a particularly hot company can negotiate a modified agreement.

Once manufacturing is complete, the vendor ships the product to the retailer and the rep gets paid.

How are rep fees structured?

Most rep groups operate on a pay-for-performance model: as the vendor ships the product to the retailer, the rep group gets paid. Commissions are negotiated and vary according to the product line and the cost of the product. "Reps have a cost of doing business," Steven says. A flight to visit a retailer, for example, may cost $300. If the product costs $10 and the vendor wants to pay only a 3 percent commission, it's going to be difficult for a rep to recoup even that one tiny piece of the cost of doing business. "The percentage that you'll pay as commission has a lot to do with the cost of your product and the potential for how many can be sold," Steven says. If a vendor can sell 10,000 units every time an order is generated, or if the replenishment rate is high, then a low commission may make sense. But for most startups, that's not the case.

What should I look for in a rep group?

Depth of relationships with targeted retailers is the most important thing. One way to judge that is to ask how much merchandise the rep group sold into the retailer. Look into the existing lines it carries and the number of placements it has in stores. Ideally, you'd like to see a rep group have dozens of active SKUs in the chains it works with; if it doesn't have enough placements, that might be a red flag that it can't get the job done. However, if the rep group has many top-tier large lines, a startup might not get the mindshare it's looking for because the account is too small. Ideally, it's best to work with a group that has the time and resources to devote to a new line getting placement.

Coverage area is important as well. While some stores have buyers who will put your product on shelves nationally, it's common for small-to-midsize retailers to be regionally focused. As a result, many rep groups focus on a specific region. "A manufacturer might have over 10 rep groups covering the US for full coverage," Chris says. Reps will often leverage their own networks to help a client achieve national coverage. A master rep can help achieve national distribution by reaching out to other reps and encouraging them to take on your product, splitting the commission.

It's important to choose the right rep group. By the time a founder is looking for one, he's often made at least a few dedicated sales hires. A good vice president of sales will be familiar with the various groups and should have pre-existing relationships. If she doesn't have connections to rep groups, you can try calling the retail buying offices and asking the buyers themselves for a referral.

Top rep groups often sell into multiple retailers, and they can provide valuable guidance on which are most likely to carry your product. They know the retailer's buying cycle. It can take many months—upwards of a year, in many cases—to successfully get onto a shelf. Reps know when to reach out and will shepherd your product through the process. They also understand common pitfalls that can arise when working across multiple channels. For example, if you choose to sell into Apple, Amazon.com, and a small specialty retailer, you may find yourself having price integrity problems. Amazon.com has a strict price-matching policy: if a third-party Amazon Marketplace seller sells your product at a discount, Amazon.com (acting as retailer) will match that price. Apple will find itself selling at a higher price, which means it will sell fewer items and that will strain the relationship. A rep group can work to manage your price and inventory across multiple stores. Or they may suggest tailoring an offering to a specific retailer. The alternative is to have dedicated salespeople within your company, each with a specific retailer relationship to manage.

One thing that surprises many entrepreneurs is the fact that a retailer's online buying team is often wholly separate from the in-store buying team. Jason Lemelson of Slam Brands explains:

> *The online programs are handled by the online buying team. In most cases, the retailers don't stock the same inventory in their stores. Retailers can carry hundreds of SKUs per category online. That's why it's typically easier to get placement on Walmart.com, Target.com, and Costco.com...they have assumed no inventory risk.*

Online buyers will verify that the vendor is able to do the warehousing, carry the inventory, and handle the pack-and-ship operation. The criteria for landing on the shelves of a physical store, where a retailer might stock no more than six products in a given category, is often far stricter. However, it's often much more lucrative. "Many times, the store buyers will redirect you to the dot-com buyers and tell you they'll put it online, use that as a proxy to determine if it's something they'd carry in the stores," Jason says. "It's a win-win proposition for the retailer, drives revenue with zero risk for them. You are responsible for essentially everything when you do the online business."

You might find logistical burdens integrating with the retailer's order-management systems. There's less analytical data than the startup would get from selling online directly, and there are many small orders to process (as opposed to one large purchase order). Full-stack ecommerce solutions such as CommerceHub might help to manage these pain points, but it can be a daunting undertaking for a young company.

Some of the larger retailers, such as Target and Lowe's, have special "innovation" divisions that are specifically set up to work with startups. Part of the innovation team's job is to identify cutting-edge products that will make the retailer seem tech-forward. Wearables and connected home devices are particularly popular. Innovation teams can help a founder in a hot sector bypass the traditional buying process. They often have the power to place a product on shelves in a single-store trial run and can work with startups to make financing challenges manageable.

Selling in a few stores as part of a pilot program is a fairly common way to move into a retail environment. For example, Best Buy reached out to Fitbit shortly after it launched. Fitbit did a pilot (*http://bit.ly/fitbt_hardware_startup*) in 4 stores, then moved into 40, and was eventually carried in 650 Best Buy locations.

Once your product is on the shelf, you face other challenges. It will be sitting on a shelf surrounded by dozens of competitive products. Even if they're not directly competitive, your product competes with them for the customer's attention and dollars on that shopping trip. Retail shelves are generally *unaided* sales; online, you can feature marketing copy directly alongside photos of the product, assisting the customer's decision to purchase. On a shelf, your packaging is all there is.

It's possible to help the customer decide to purchase your product by reaching out to the retailer's employees. Eric Klein emphasizes the importance of making friends with the "blue shirts" (a reference to the blue shirts worn by Best Buy employees): "If they don't like your product, you aren't going to sell." If there are three related products sitting on a shelf—say, two made by big brands and one made by you—a customer will often ask the salesperson for help deciding which one to buy. If the salesperson has no idea what your product does, or doesn't like it, you're at a real disadvantage. "At Dash, we allocated part of our budget to making sure that every employee at the company, from Quality Assurance Engineers to the CEO, went out into the field on the weekend, sat by the shelf, talked about the product, and met the blue shirts," Eric says.

Besides winning the hearts of the salespeople, another strategy is to hire *detailers*: companies whose job it is to go in and check product placement on shelves, make sure the products are tidy and organized, that boxes are intact. Sometimes a detailer will man an endcap on an important weekend and talk about your product to customers who walk by, or might set up a demo table. Detailers can be incredibly expensive to use, so this strategy is typically employed by larger brands.

To facilitate better treatment and placement, most companies that sell in big-box stores budget for MDF: extra money paid to a retailer for prime features, including endcaps or banners. Coupons, promotions, and placement in circulars also fall under this budget. The company (in this case, your startup) eats most of the entire cost of a promotion. Some retailers require certain types of promotions at various times of the year, such as holidays or back-to-school shopping.

Theoretically, the lift (increased sales) offsets the discount, but that isn't always the case. "In some cases, we'd see lifts in the range of 20x from circular placements," Jason Lemelson says. "Some of them, such as the holiday circular at Toys R Us, are so effective that the retailers charge a fee for inclusion." An ad in the Costco Connection can run as high as $90,000. Some retailers negotiate a coupon schedule before they agree to place the SKU. Jason adds, "The ad may cost tens of thousands, but supporting a $30 off coupon can run up into the millions." MDF budgets vary, but many companies think of them in terms of the percentage cost of the item—e.g., 45 percent for retail margin, another 3–7 percent for retail MDF.

The final financial hurdle for a retail distribution strategy is getting to the point where you can meet the inventory demands of retailers who are offering net-30, 60, or 90 payment terms, and still keep the lights on. Many larger companies use what's known as *factor financing*.

Factor financing, or *factoring*, is a financing method that allows a company to have access to cash several months before the Net D (D is the number of days) payments are due. It isn't a loan; instead, the company sells a future cash flow to a third party at a discount, in exchange for more timely access to capital. Let's say you've received a purchase order from a large retailer. The purchase order is typically for a large quantity of units, which would require a substantial capital outlay. A third-party financial services company (called a *factor*) buys that receivable from you, often for several percent below face value. In exchange, they pay you a portion of the cash immediately, typically up to 85 percent of the value of the purchase order. For example, if you have a $1 million order from Target that you sell to a factor, the factor would advance you $850,000 for it. You turn the receivable over to them. The factor informs Target that it now holds the invoice, and Target pays the factor. Once Target has paid the $1 million, the factor sends you the balance less its fee.

Factoring rates depend on several considerations, such as how well-established your company is, how strong your business looks, and how credit-worthy your customer is. CIT and Wells Fargo are two of the larger financial services firms that offer the service. Because startups are rarely pulling in tens of millions of dollars of receivables, CIT and Wells Fargo are often not interested in working with them. Some boutique banks, such as Silicon Valley Bank, may be more amenable. While there are other providers of trade financing, boutique factor rates may initially be more costly than bank loans. But it often takes less time to secure factor financing than to secure a loan, factor financing rates can improve over time (versus the fixed interest rate of a loan), and even relatively new companies can qualify for good rates provided that their customers have solid credit. It's important to consider all options carefully.

Retail is extremely complex, and it's vital to weigh the pros and cons carefully before starting down the path. In his blog post "Why Consumer Hardware Start-ups Fail" (http://www.inc.com/barros/why-consumer-hardware-start-ups-fail.html), Marc Barros, founder of Contour Cameras and Moment, emphasizes the importance of putting retail distribution in its proper strategic place:

> The lure of selling in a megastore was so strong, we [Contour] started preparing for distribution before we even had a product to sell. We wound up spending our limited resources on product and distribution, leaving nothing to become the brand of choice. Our competition took the opposite approach. By selling almost 50 percent of their product consumer direct, their high margins allowed them to pump significantly more back into marketing. Their ability to reach new customers continued to snowball, creating an umbrella of awareness we could never penetrate.

If you choose to go this route, it's critical to know exactly what you're getting into before you jump, and to have the capital and headcount available to make a retail strategy work for you. Don't be fooled into thinking that your product will sell itself as soon as it's on the shelf of a big-box store. As Marc puts it, "Distribution doesn't increase customer awareness. It only fulfills orders for the demand you've created."

WAREHOUSING AND FULFILLMENT

We've talked about distribution channels; now let's talk about the logistics involved in distribution. If you're selling direct, there are three options for handling fulfillment. First, you can either hold the goods in your own warehousing facility (this can be your garage, in the early days) and ship from there, or you can hire a third-party provider to handle storage and shipment for you. Sometimes a factory might be equipped to handle shipping directly.

If you're selling in an etailer, both of these models are possible options. The third option is to have the etailer take inventory and shipping. That model is most common when the etailer is retailing your product rather than just providing you with an online platform. Amazon Marketplace is a notable exception; it offers the Fulfilled By Amazon program (discussed later in this chapter).

Regardless of whether you're planning to implement self-managed or outsourced fulfillment, an order management system (OMS) is key. Out-of-the-box ecommerce solutions such as Shopify (*http://www.shopify.com/*) and Volusion (*http://www.volusion.com/*) will keep track of orders, collect payments from customers, provide shipping labels, and manage inventory. More specialized OMS platforms offer demand forecasting and vendor management, which can help you plan for how much to manufacture in each run. Inventory management is an exercise in optimization: your goal is to minimize storage costs and avoid tying up capital, while still producing enough units to get volume discounts, and having enough stock on hand to meet customer demand. If you're doing a crowdfunding campaign, you can stagger fulfillment dates: have one reward tier for a first run that ships in December and then a slightly lower-priced tier for a run shipping in February. You'll know exactly what you'll need, and when. But things get more complicated when you're planning for regular operations. Tools such as Lettuce (*http://bit.ly/lettuce_tool*) offer predictive models that can help.

If you choose to handle your own warehousing and shipping operation, location is key. Liam Casey, CEO of PCH International, believes that an ideal manufacturing and fulfillment strategy places your team no farther than three hours from your factory, and the goods no farther than three days' time from a majority of your customers. If you're in an urban area, such as the San Francisco Bay Area or New York City—and most hardware startups are—rent will be extremely costly and space is often difficult to come by. It can take several months to find an optimal space, particularly one that's easily accessible by truck and equipped to receive and store pallets.

Alex Andon, founder of Jellyfish Art (see "Jellyfish Art: A Case Study" on page 263 to learn how Alex built an elaborate warehouse operation for storing—and breeding!—live jellyfish), notes, "If you need under 2,500 square feet, you're likely to have to sublease. Rent in the San Francisco Bay Area is a dollar a square foot per month and up." When you're budgeting, don't forget to consider garbage removal, insurance, security, WiFi, telephone, water, and electricity costs. You will also be paying wages to the people who move, pack, and ship the boxes. Alex emphasizes that order fulfillment is about planning ahead and staying organized. "It's important to balance cost, control, and convenience."

Jellyfish Art: A Case Study

Alex Andon is the founder of Jellyfish Art. His company makes a beautiful tank for jellyfish, which are notoriously difficult to care for in standard filtered home aquariums. In addition to selling the tanks, Jellyfish Art breeds and ships the live jellyfish...which left Alex with some particularly complex challenges around warehousing and fulfillment.

Long fascinated by public-aquarium jellyfish exhibits, Alex started experimenting with making tanks in his garage. After interest from family and friends, he started selling them online. As demand grew, he decided to put them on Kickstarter to gauge market demand. The stated Kickstarter target was $3,000. By the end of the campaign, he'd raised $163,000. Blogs, television, and radio stations featured the project. What he had thought of as a way to generate an extra little bump in sales made him realize that his project had the potential to be a business.

He took the standard approach to scaling up manufacturing: he went to China and found a factory that would produce the tanks. But then he had to figure out how he was going to get those tanks—and their live inhabitants—to customers. Since Jellyfish Art was now a full-time endeavor, the solution had to be scalable. Alex recalls:

> Before the Kickstarter, it was me and my cousin just filling website orders from a garage with a few holding tanks that we had put together for the jellyfish, which at that point we were getting from a supplier. All of a sudden, we had to jump up to a huge warehouse, hire multiple employees, find pallet racks, thousands of gallons of holding tank space...we were building up all of that infrastructure while our factory in China was building the tanks.

The team decided to breed their own jellyfish to reduce supplier costs and offer customers more options. This required a temperature-controlled space with electricity that was big enough to house the tanks. Warehouse space under 1,500 square feet is difficult to come across. At that small size, it's often necessary to sublet. Jellyfish Art wanted to avoid that, so when a 5,000-square-foot space opened up nearby, they jumped on it, subletting the portion they didn't yet need.

Simultaneously, the company was coming to the conclusion that dropshipping the tanks from China was going to be an expensive undertaking. Daily orders were coming in, and each tank weighed 20 pounds. Air freight costs from overseas were simply too high. The team decided that both halves of their fulfillment process could best be managed in their own warehouse. They calculated their approximate inventory needs for a year and began to do large runs, which took three months of manufacturing time. Alex explains:

The product itself has about 10 components that come from other factories, but our point of contact in China is the factory that produces the acrylic tanks. They bring in components from the other five factories and assemble it all. They are also responsible for getting it to the shipping company in China, so they load it on a truck. Once it gets from the factory to [Shenzhen], FedEx takes over; they're responsible for getting it on the boat to San Francisco, and then getting it from there to our warehouse. The boat shipments take about a month to arrive.

Shipping within China is inexpensive, as the bulk of the shipping cost is the container shipping to the US. To economize, Alex orders the tanks in units of container loads. He's aware of inventory risk and never orders more than he anticipates selling in a year. He also employs the services of a customs broker to make sure that all of the necessary paperwork is filed in advance of the tanks getting on the boat:

FedEx assumes you know what goes into the customs process, but most first-time entrepreneurs don't. The advantage of a customs broker is that they're an easily reachable point of contact. They are relatively inexpensive and make everything much easier.

Now, the Jellyfish Art warehouse has pallets of tanks on risers, and jellyfish-filled breeding aquariums filling out the ground floor. When an order comes in, the team simply pulls a tank off of a pallet, puts a shipping label on it, and mails it out. After the customer has received and set up the aquarium, he returns to Jellyfish's site and submits a code to receive the live shipment. The translucent jellyfish are packaged with great care and sent off to their new homes.

In an outsourced model, fulfillment companies offer warehousing, packing, and shipping services. Fulfilled By Amazon is the largest player in the outsourced fulfillment space. There are also smaller companies and startups, such as Shipwire (https://www.shipwire.com/) or Whiplash Logistics (https://www.whiplashmerch.com/).

Fulfillment companies typically charge a monthly fee for storage plus a per-order fee. The order fee includes pick fees, pack fees, and postage. Shipwire and Whiplash both feature calculators on their site, to help customers understand the storage and handling fees.

The number of products and size of boxes impacts the amount of space you'll need, while the number of SKUs increases the complexity of the packing process and, as a result, the cost. For example, using Shipwire, two 2-SKU company that has 100 orders a month averaging 1.4 items per box and requires two pallets of goods stored in Los Angeles will pay $353.95 per month plus $2.89 per order handled—not including postage.

Amazon Fulfillment has low rates: $1.00 per customer order, $1.02 pick and pack per unit for electronics, plus variable weight-handling and storage costs. The primary reason that customers of smaller logistics providers cite as the reason they did not go with Amazon is personalization and branding. Fulfilled By Amazon will ship your product in an Amazon box with Amazon papers inside. A smaller company can offer more personalized customer service and can wrap and pack your product in whatever specific manner you choose.

James Marks, founder of Whiplash Logistics, notes that the decision to move from DIY fulfillment to an external service often requires a formalization of the process, "If someone is shipping 5 or 10 orders a week, shipping isn't really a problem that needs to be solved." If you're shipping small quantities in house, you can still be somewhat poorly organized because there's in house knowledge about how the product should be shipped. "Once you go to an external center, everything must be specified," James says. "You have to have SKUs, you have to start thinking about bar codes, you need to become much more systemic and organized."

When you're deciding which service to use, location is an important consideration. You want to be either close to your customers, if your customer base has strong regional demographics, or close to your factory or office. "People feel better knowing that they can go over and see the warehouse," James Marks says. A tech-savvy partner is also desirable. He points out, "Some of the older-school companies don't plug into online shopping carts."

The shipping rates that you see at the post office are not what you should be paying if you're doing meaningful volume. FedEx, UPS, DHL, and USPS will all negotiate volume shipping discounts. "Get a sales rep and lock in rates early," Alex Andon says. "DHL is the most cost-effective for international shipments. USPS is the cheapest for small boxes. FedEx and UPS are general purpose. Pick one and stick with it." Consider using a shipping service API such as Shippo (*https://goshippo.com*) or EasyPost (*https://www.easypost.com/*) to take advantage of lower rates and to see price comparison across carriers. If you're using an external service, such as Whiplash, the fulfillment company may pass down its volume-shipping discount to you.

For brick-and-mortar retail, etail, and outsourced fulfillment, you are responsible for getting your product from the factory to the door of the warehouse or distribution center. If you are selling in any type of brick-and-mortar store, the store holds the inventory. It takes responsibility once the inventory hits the warehouse or distribution center, and it will handle shipping, returns, and customer service for those consumer touchpoints.

If you're fulfilling orders yourself, you'll be handling customer service yourself. Basic customer service practices include keeping your customers updated on their order progress, providing them with email confirmations and a shipment tracking number, and posting an easy-to-find customer service email address and/or phone number. If you are a small team, it might be difficult to do more than this in the beginning.

To enable small teams to provide exceptional customer service despite limited resources, Alex Andon suggests creating an online knowledge base containing instructional videos for customers. These are particularly useful for potentially common issues such as getting started or initial configuration. Analytics attached to these assets can give you insight into common customer problems as you see what people are watching or searching for. This can help shape the next version of the product. Either way, it will help you keep your early customers happy.

Many business books have been devoted solely to the concepts briefly presented in this chapter. Pricing, positioning, marketing, and distribution are all extremely complex. We decided to provide an overview that is comprehensive enough to give a founder a sense of how to approach these problems.

CHAPTER 11

Legal

THIS CHAPTER PROVIDES A BROAD OVERVIEW OF SOME OF THE LEGAL CONsiderations that hardware startups are likely to encounter. Writing about legal considerations is challenging; there's no one-size-fits-all strategy that applies to every hardware startup. However, issues such as regulatory compliance, patents, and industry certifications are highly likely to come up as you build your company and sell your product. The material presented here is meant to be a broad overview of the big topics that a founder needs to be aware of. It isn't legal advice or a substitute for talking to qualified lawyers about your specific situation.

There are six areas of legal complexity that almost every hardware company will encounter:

- Company formation and establishment of a legal entity
- Intellectual property protection
- Contracts (for sourcing, manufacturing, and the sale of finished goods)
- Liability concerns
- Regulatory considerations
- Industry certifications

One lawyer is unlikely to have expertise in all of these areas, so hardware startup founders commonly work with multiple types of lawyer. In the early days, a corporate lawyer will help with entity formation and contracts and can often provide advice about how to handle potential liability issues related to the product. However, most corporate lawyers don't have extensive experience with patents, so a patent attorney is often needed as well. She can help you navigate complex intellectual property issues. If you're working within a highly regulated industry—say, building a medical device—you might want to hire a lawyer or consultant who has taken many such devices through the approval process.

Company Formation

This process is essentially the same for hardware and software companies. The primary decisions to make are what type of legal entity to choose and where to form the entity. It isn't necessary to form a legal entity as soon as you have an idea. Some founders wait until they've determined that their idea is viable and form a legal entity after the customer-research phase. However, it's typically best to form the entity prior to prototyping and manufacturing, particularly if you have cofounders. You should certainly do so before hiring employees (especially if you plan to award stock options), and/or before taking outside funding.

Founders should be careful about entering into significant agreements prior to forming a business entity, according to Josh Fisher of Bay Area firm Fisher Law Offices (http://jfisherlaw.com/). "Even if the agreement is made in the name of the business entity," Josh says, "the founder will generally be liable personally for any related obligations unless and until the business entity is formed, the entity takes over the agreement, and all parties to the agreement release the founder."

Startups in the US typically organize as one of three types of legal entity:

C-Corporation
A C-Corporation is a legal entity that is subject to federal income tax. This structure is what most people tend to think of when they envision a "corporation." Most publicly listed companies on stock exchanges are C-Corps. Taxation is handled at the corporate level rather than passed through, though individual shareholders do pay taxes on any distributions or dividends they receive. Generally speaking, if you are planning to grow your business into a large company and raise money from outside investors, this is the optimal choice. A C-Corp's shareholders are protected against personal liability (beyond any investment they've made in exchange for their shares) for the corporation's obligations. This is known as *limited liability*. There is no limit to the number of shareholders, and C-Corps are not subject to any citizenship or residency requirements.

LLC

LLC stands for *limited liability company*. This structure is typically used by smaller businesses that don't plan to raise outside funding. It has a lot of flexibility, particularly in terms of ownership structure and tax treatment. LLCs can be held by sole proprietors (or a married couple holding as a sole proprietorship). Unless they elect to be taxed as C-Corps, LLCs are considered *pass-through* entities for tax purposes, meaning that profits and losses are reported on the proprietor's personal income tax. This is known as *pass-through taxation*, because profits are not taxed at the LLC level; instead, they pass through directly to the owners' tax returns. If the LLC is held by a partnership, it files a separate tax return but taxes are still paid through the owners' personal income taxes. This can occasionally become complex. If the business is active in multiple states, participants might be obligated to pay taxes in the active states even if they don't personally live there. An LLC also provides its owners with limited liability.

S-Corporation

S-Corporations are similar to LLCs in that shareholders benefit from limited liability and pass-through taxation. However, an S-Corp requires that the owner pay himself a salary for managing the business. S-Corps are limited to one class of stock and 100 shareholders, who have to be US citizens or residents and *natural persons* (real people, not partnerships or corporations). Be aware that this limitation on the number of shareholders can prove challenging if you intend to award stock or stock options to employees.

If you are considering doing a venture capital raise or taking outside investment, be aware that most investors prefer a C-Corp structure. Most institutional venture funds can't easily invest in LLCs because of their own legal structure or tax considerations of their limited partners (if you're interested in learning about the specifics of why, a blog post by Ryan Roberts (*http://bit.ly/c_corps_for_vc*) provides a clear explanation). And since VC firms frequently want preferred stock, the single-class structure and natural persons requirement of an S-Corp are typically dealbreakers. It is possible to convert your company from an LLC to a C-Corp, but the process can be costly and time consuming.

After you have decided what type of legal entity to be, you have to decide where the entity should be formed. Where to form largely depends on where you are likely to do the most business. Most legal assistance sites for businesses will tell you that if you are going to do a majority of your business in one state, you should form your legal entity in that state. That advice is typically intended for people planning to open a business serving the local community, or a mom-and-pop shop. If you're starting a hardware company that will be selling a product and will likely have customers across the US, you might also want to consider forming your entity (http://bit.ly/where_2_incorporate) in Delaware.

In addition to where you'll do the most business, these other considerations should figure into determining where to incorporate:

- Tax rates (franchise tax)
- Costs of formation in a given state versus costs of registering as a *foreign business entity* in that state*
- Business entity/corporate laws of the state with regard to rights of creditors
- Laws of the state concerning the rights and responsibilities of managers and shareholders

For most early-stage startups, the last two are largely irrelevant. They become relevant only during later-stage equity financing rounds and as your company approaches initial public offering (IPO). Protection from derivative lawsuits (http://bit.ly/incorporate_in_ca_or_de) (suits initiated by shareholders on behalf of the entity, usually against a corporate executive) is also stronger in Delaware. This is why approximately half of all of the corporations listed on the New York Stock Exchange are Delaware corporations, even though the majority of their business happens elsewhere.

Startups do benefit from many aspects of forming their business entities in Delaware. Delaware has low formation fees and no corporate income tax for companies that are registered in Delaware but don't have operations there. There is no requirement for a physical corporate presence in the state, and no name or address disclosures are required for the initial board of directors (though if you have no physical presence in Delaware, you'll have to engage a Delaware-based registered agent). As a result of having so many businesses incorporated in the state, incorpora-

---

* A *foreign business entity* is a corporation doing business in a state other than its state of incorporation.

tion processes tend to move quickly, and the state offers rush services for a number of forms. Delaware also has a court system called the *Chancery Court* that uses judges (rather than juries) who have extensive expertise resolving complex business disputes.

If, like the majority of hardware startups, you are based in California (*http://bit.ly/de_corp_in_ca*), forming your entity in Delaware will require you to pay for a Delaware-based registered agent as well as franchise taxes for both states. And, as mentioned earlier, you must register as a foreign business entity within California.

Some VC firms and accelerator programs, such as Y Combinator, actually require that a company be a Delaware C-Corp. Although most YC companies file incorporation papers while participating in the program in the San Francisco Bay Area, the accelerator (and many VC firms) prefers Delaware because of the protections described earlier. There are also legal considerations (*http://bit.ly/incorporate_what_state*) surrounding mergers and acquisitions. In California, a majority of outstanding shares across all share classes must approve a merger, while in Delaware only a majority of outstanding shares entitled to vote are needed.

There are many great resources that cover the finer points of entity formation. While no blog can replace having a firsthand conversation with a qualified corporate lawyer, two excellent sites featuring plain-English resources on corporate structure and incorporation advice for startups are (similarly named) Startup Company Lawyer (*http://bit.ly/incorporation_reqs*) and Startup Lawyer (*http://startuplawyer.com/*).

For bootstrapping entrepreneurs who want to file their own entity formation paperwork, it is often possible to do so by faxing the necessary documents to the state. There are also startups devoted to making the process more DIY and less expensive, such as LegalZoom (*http://www.legalzoom.com/*), Docracy (*https://www.docracy.com/*), and Clerky (*https://www.clerky.com/*).

Some lawyers will do your incorporation paperwork pro bono prior to a venture capital raise. This is particularly common in Silicon Valley, where firms frequently work with startups. Several, such as Orrick (*http://www.orrick.com/*) and Fenwick & West (*http://bit.ly/fenwick_and_west*), offer value-added services that include investor introductions, lectures, bootcamps, and founder networking. In New York, Cooley has a program for hardware entrepreneurs.

> **TIP** Briefly departing from lawyers for a second here: you'll also need a company bank account, and a number of banks offer value-added programs similar to the law firms just mentioned. Silicon Valley Bank (*http://www.svb.com/accelerator/*) and First Republic Bank (*https://www.firstrepublic.com/business*) are two in the Bay Area.

While finding a good lawyer may seem daunting, many related discussions threads on Quora get into city-by-city specifics. Some firms that appear repeatedly include Goodwin Procter (*http://www.goodwinprocter.com/*), Gunderson Dettmer (*http://www.gunder.com/*), Lowenstein Sandler (*http://www.lowenstein.com/*), and WSGR (*http://bit.ly/wsgr_firm*).

And just to reiterate: reading about incorporation and taxation issues in a book isn't a substitute for consulting with a qualified attorney and/or certified public accountant (CPA). Your startup's circumstances may make a certain type of structure more advantageous for you than it typically is for the majority of founders.

Trademarks

A trademark (*http://bit.ly/tm_basics*) is a symbol, word, device (sound/color/smell), or name that is used to identify the seller or provider of a particular good or service. Trademarks help you to distinguish your product from a competitor; they're an important component of brand identity. Within the US, trademarks can be registered with the Patent and Trademark Office (USPTO) and/or with the trademark offices in the relevant states. The US system is based on first use (*http://bit.ly/foreign_ip_strategies*): whoever uses the trademark in commerce first generally has the best claim to it.

Many other countries, including China, take a first-to-file approach to trademarks. David Pendergast, vice president of corporate development and legal affairs at PCH, advises startups to be proactive about filing for their trademarks in China *before* they go public with their brand, and certainly before a crowdfunding campaign. "China has been putting more and more emphasis on enforcing intellectual property rights, which is a good thing," he notes:

> *In China, there is no trademark protection until you file. In this case, first-to-file wins. For example, if a startup gets a lot of press, has a crowdfunding campaign, or gains notoriety or success, there is a strong likelihood that someone in China will see that and file the trademark in China to gain the*

rights. This becomes a problem when export authorities compare the trademark on product packaging with who owns the Chinese trademark on the trademark registry. If they are not the same person, the shipment will be blocked. The only way to solve this discrepancy is to pay the trademark squatter to sell the trademark in China. Thus, we strongly recommend that companies secure the trademark in China before engaging in press, crowdfunding, or promotional activity.

Even if you're not planning to manufacture or sell your products abroad, protecting your trademark by filing both domestically and overseas is an important thing to do. Among other benefits and protections, the US Customs and Border Protection Bureau of the Department of Homeland Security can help prevent importation of counterfeit and grey-market goods into the US when the goods violate your federally registered trademark. Your trademark is an important asset when you communicate with your customers. Preventing a squatter from claiming it in the first place is easier than fighting one who already has it. For more information on trademark protection, see the US Patent and Trademark Office guides (http://bit.ly/tm_squatters) on the subject.

Trade Secrets

Generally speaking, a *trade secret* is any information that has economic value to your business, that isn't publicly known or easy to figure out, and that you have used reasonable efforts to keep secret. Trade secrets are protected by law against theft and misuse (known as *misappropriation*) by third parties. Josh Fisher says, "Protecting your trade secrets can involve a variety of measures, but a basic one is to ensure that all of the founders and any employees or independent contractors sign non-disclosure agreements covering your company's confidential information."

Patents

A patent (http://www.uspto.gov/patents/) is an intellectual property right granted to an inventor "to exclude others from making, using, offering for sale, or selling the invention" in a given domain (such as the US, if the patent is awarded by the US government).

Deciding whether or not to pursue patents is something of a debate among hardware startup founders. In his book *From Concept to Consumer* (FT Press), product design expert Phil Baker describes his experience with the patent process, beginning with his time as a product developer at Polaroid. At Polaroid, he was encouraged to patent processes that the company used, as well as to defensively patent ideas that were shelved. The latter patents could still be used to prevent competitors such as Kodak from encroaching on Polaroid's market leadership in instant photographic development.

While this strategy served Polaroid well, Phil's subsequent experience bringing consumer products to market changed his opinion on the matter. "Much has changed in today's consumer world, where products are developed in months instead of years and can last less than a year in the market rather than several," he writes. Development happens increasingly rapidly, but it still takes years for a patent to be granted. Because of this disconnect, Phil believes that pursuing patents can detract from a team's focus and lead them to spend money unnecessarily. He writes:

> *With shorter development times, a company can often introduce a competing product in months, years before your patent will even be issued! By the time your patent does issue, both you and your competitor's products may no longer even be on the market.*
>
> *Even when your product is granted, it's rare that you can stop an infringing company from selling its product. You'll have to challenge the company in court, and that can take years and cost hundreds of thousands of dollars. Your only recourse is to obtain an injunction to halt the company's sales, but injunctions are not easily granted.*

On the flip side, filed patents can signal to a larger company (or patent troll) that a startup is not to be messed with. In 2012, large incumbent Honeywell sued Nest when it entered the smart-thermostat market. The complaint accused Nest of infringing upon seven patents related to operating and setting thermostats and included a requirement that Best Buy cease selling Nest's product. Honeywell issued a press release (*http://bit.ly/honeywell_nest_suit*) announcing the lawsuit:

> *The patents are related to, among other things, simplified methods for operating and programming a thermostat including the use of natural language, user interfaces that facilitate programming and energy savings, a thermostat's inner design, an electric circuit used to divert power from the user's home electrical system to provide power to a thermostat, and controlling a thermostat with information stored in a remote location.*

Nest, which at the time was a young startup that hadn't pursued a patent strategy, responded with a counterclaim (*http://bit.ly/nest_honeywell_counter*), accusing Honeywell of attempting to inhibit competition and disingenuously failing to disclose even its own prior art for several of its patents. A year later, Nest entered into a patent licensing agreement with Intellectual Ventures. Considered by many to be a patent troll, Intellectual Ventures owns nearly 40,000 patents, purchased from inventors and then licensed or sold to companies, particularly those that might need to defend themselves against patent disputes by using patents that the aggressors themselves are conceivably violating.

San Francisco patent attorney Jeffrey Schox calls this type of strategy "Mutually Assured Destruction" (see "Patenting Tips from a Pro" on page 277). His boutique firm, Schox Patent Group (*http://www.schox.com/*), helps many hardware startups develop a patent strategy. According to Jeff, there are three primary reasons why startups should consider filing patents, and they aren't the same as the reasons that make patents appealing for larger companies. Larger companies can enforce their patent rights by filing patent infringement lawsuits; smaller companies rarely have the resources to pursue such a strategy. Instead, a startup should build a patent portfolio to increase valuation (either in a fundraising round or during an acquisition), deter patent lawsuits from large competitors, and increase leverage during negotiations with a strategic partner.

Patenting Tips from a Pro

In this Q&A, San Francisco patent attorney Jeffrey Schox offers insights into the patenting process.

When is the right time for a hardware startup to pursue a patent?

A lot of founders don't get the timing right. There are two competing factors: knowing what's special and getting an early filing date. Founders want to protect the thing that they believe is really unique; usually that's the hardware, and they know that patents and hardware go hand in hand. But there are so many companies out there that it's rare for hardware to be unique. We want founders to be patient, to understand what's really special about their offering. What is their real differentiator, and what is the technology behind that differentiator? Being patient is at odds with the first-to-file patent system. You get your date, you beat other people to filing, by getting to the patent office early. We tell our clients that it's important to learn what really is special, but it's also important to file before launch.

Given that, when should a founding team first seek you out?

We love to hear the story early, to be able to help lay the foundation for when we actually launch the patent application process. There's an onboarding process to get to know a new client. If someone shows up a week before launch, there's probably very little we could do to help them. Typically, patent firms have a two-month queue of work.

Does every company looking to build a physical device need to have some kind of patent?

There are folks that don't have or need patents. I think it depends a lot on the competitive landscape—who you're going up against—because deterring patent infringement lawsuits is something that's particularly important. One of the unfortunate realities right now is that with other companies having and building patent portfolios, the best defense against someone else suing you is to actually have your own patents. This deters patent litigation. Large companies don't want to sue smaller companies, if they have patents that [the larger company] might themselves be infringing. They actually have more to lose in that lawsuit than they have to gain by shooting down a smaller startup.

We view that as one of the three major reasons to file: deterring patent infringement lawsuits by a competitor. The other reasons are to build both short-term and long-term valuation and to protect or increase leverage with a strategic partner. It's important to the long-term success of the company, and filing early is one of the only ways of actually getting those patents in the first place.

Should founders do a patent search before they start to develop their own product?

This sounds a little counterintuitive, but if you're an expert in the space and you have an interesting take on something, just go for it. Since we are unable to search the patent applications filed within the last 18 months, it is often a waste of time and money to conduct a thorough patent search.

How do you think about patenting for an IoT or wearables company with a significant software component?

We might file two patents. Typically, we file a patent application that has the whole ecosystem from hardware to cloud: the big data, analytics, algorithms aspects in it. The focus may be more on the software. And then we may also file another application that focuses more on the sensor, if there's something particularly interesting about the sensor or the hardware itself.

What's your advice for a startup that's competing in a crowded space? For example, the technology isn't new, but its differentiator is price, or it's focusing on building a brand targeting a particular type of consumer.

At a minimum, filing a design patent is possible. They're inexpensive and easy. It says to the world, "We've invested a lot in design; don't copy it." This works for a product that has some uniqueness to it—a particular cut, a particular angle. Something that's simple and elegant can be tough.

There are certainly startups that come to us and we end up having an hour-long brainstorming session and then just throw up our arms, saying, "There's really nothing here to patent."

What are the typical costs a team should expect with a design versus a utility patent?

A design patent might just be a few thousand dollars for all things included—the drawing fees, the government fees, our work—and it can often be done in a couple of days after the onboarding. A provisional application for a utility patent is slightly more, but usually a few thousand as well. We charge $5,500; again, that's with legal fees, the drawing fees, government fees all included. And a typical utility patent application is within the $12,000 to $20,000 range.

Where do startups make mistakes when thinking about patents?

I think a lot of times founders skip over the provisional process. They've heard from someone that provisionals are worthless or that anyone can get a provisional. If you ask a patent firm to write a 20-page provisional application, there's a very, very high likelihood that you're going to get that date. And you're only going to spend a fraction of the money that you would on a full utility patent application.

That provisional will give you 12 months in which you can continue to evolve and work on the implementation. The full application often includes about two or three times as much information as a provisional. That depth, having those deep implementation details, dramatically increases the chances that we're going to walk away from the patent office with an issued patent in hand.

Is there anything to be gained by patenting overseas—for example, before reaching out to a contract manufacturer?

We do file a decent number of patents in China. The main reason is to protect against manufacturers pulling another shift. It's still yet to be seen how effective those will be. There are more patent applications filed there [in China] than any other patent system in the world, and they are becoming more and

> more aggressive with respect to protecting intellectual property rights. This is a policy matter. I believe that they'll flip the switch at some point, when they believe that they no longer need to copy the technology of the rest of the world. I think we'll be filing an increasing number of patents there going forward. It makes some sense to patent in Europe as well.

Whether you decide to patent or not, knowing *what* is patentable is important. Utility patents (*http://www.uspto.gov/patents/process/*) can cover either a device, material, or process. It's also possible to file a design patent that covers *ornamental characteristics* of a product. The patent office requires a patentable invention to be novel and nonobvious. With the right degree of specificity, it might be possible to get any number of devices or methods patented, but such patents might not be valuable. Too-narrow applications not only waste founder time and money, but they can also teach competitors how your invention works.

Jeff advocates patenting core technology and the functionality that it enables: things that are *both patentable and valuable*.

Should you decide to pursue patents, the process typically involves working with a patent attorney to file first a provisional application, then a full patent, and finally, to respond to the patent office examiner's concerns. The purpose of the provisional application is to establish an early effective filing date, potentially preserving patentability. The USPTO charges $130 for the filing (as of April 2015). Jeff estimates the legal fees for the provisional application to run between $3,000 and $6,000.

The conversion of a provisional into a full patent application must happen within one year. At that time, the inventor files detailed specifications, drawings, and other materials describing the invention, makes formal *patent claims*, and submits an oath stating that she believes she is the original inventor. The USPTO provides extensive descriptions of the requirements (*http://www.uspto.gov/patents/resources/types/utility.jsp*) and flowcharts covering the process. The USPTO charges $800 for the filing fee. Jeff estimates that the lawyers' involvement in the process takes six to eight weeks (encompassing 30–40 hours of attorney time) and costs between $9,000 and $18,000.

It often takes two to three years before the USPTO examiner evaluates the patent. An expedited *prioritized examination* option (called *Track One*) is available for an additional fee. Once the examiner examines the application, there is often a series of examiner rejections (often for reasons related to *prior art*) and responses by the inventor. Each rejection and response may add several months to the process and cost additional legal fees, but this work is typically done by the lawyer and doesn't take up significant amounts of a founder's time. All fees and expenses included, Jeff Schox estimates the total cost of the process to be in the $20,000–$40,000 range, spread over a two- to five-year period.

Patent law and the filing process are both extremely complex. If you're interested in learning more, Jeff helpfully walks entrepreneurs through the patenting process in an accessible way in his book, *Not So Obvious: An Introduction to Patent Law and Strategy* (http://bit.ly/not_so_obvious) (CreateSpace).

Manufacturing Concerns

Working with a factory or supplier for the first time can feel a bit daunting. Many founders are worried about not getting what they paid for, or encountering long and costly delays, or having their IP stolen. There are several legal areas for entrepreneurs to be aware of.

LIABILITY

Liability protection is an important consideration for hardware founders. If something goes haywire with your product and results in a personal injury or property damage, you want to be sure that you have enough protection in place to deal with any financial repercussions. Product liability insurance can help to protect a business from claims related to manufacturing flaws, design defects, and failure to provide adequate instructions or warnings. While general commercial insurance may cover some of these types of claims, it's important to consult with an insurance professional to make sure that your company is adequately covered. David Pendergast of PCH advises getting appropriate indemnification from the manufacturer and any external design teams as well.

Being organized as a corporation or LLC shields the founding team and shareholders' personal assets in almost all situations. As with all information in this chapter, be sure to verify with your lawyer that this is the case for you. There are often state-specific corporate insurance requirements.

MANUFACTURING AGREEMENTS

A *manufacturing agreement* is a legal contract that specifies the products and services your manufacturer is going to provide. The specifics vary by product, order size, and size of the contracting company. Larger companies often include more provisions than startups. These agreements may include:

- Tooling
- Engineering and/or design services
- Minimum order quantity
- Ownership of intellectual property and profection of confidential information rights
- BOM cost (broken down into component parts, potentially including packaging)
- Quality specifications
- Provisions for handling defectives (both in the factory and after purchase)
- Provisions for inspection of the goods
- Consequences of a recall
- Payment dates and terms
- Indemnities for IP infringement, personal injury, and property damage

Some larger companies include the right to terminate the contract or reject the shipment if a certain percentage of defects is exceeded. Most are also concerned with factory capacity, though this typically isn't an issue for hardware startups. You'll find many helpful sample contracts (*http://bit.ly/contract_mfg_agreemnt*) online.

It's important to have a contract in place, but David Pendergast of PCH cautions entrepreneurs:

> *Your relationship with your manufacturing partner is more important than what the contract says. In my view, there's a lot of time wasted on putting operations details, which really don't need to be in a contract, into the contract. If you have a solid relationship with your manufacturer, they will focus on results, not on the contract details. In the case of problems, a good relationship means you and your manufacturer will have a conversa-*

tion about the problem and work to reach a solution. Excessive worrying about what's defined in the contract or inserting numerous "prior written approval of the client" requirements will necessarily slow things down and possibly be an impediment to solving real problems.

In the early stages of your company, it's most important to find a trustworthy manufacturer that will work to make things right if there is a problem. You can't foresee—or iron out—every conceivable issue in a contract.

IMPORT/EXPORT CONSIDERATIONS

Chapter 10 briefly touched on importation issues. If you're importing goods or supplies from a foreign country, you'll need to be aware of customs requirements, tariffs, and fees that may apply. If you're exporting goods, you'll need to be aware of the US export control regime, which is currently implemented by the US departments of State, Commerce, and the Treasury. (See export.gov for details.)

Depending on the freight professional you hire, customs requirements, tariffs, and fees might be dealt with for you; if not, you'll need to familiarize yourself with the ones that apply to the region you manufacture in and the type of product you're making. Some import fees are specific to a particular mode of transportation (air, ocean), or a particular class of goods, so this section covers the ones that are most common. This list might not be comprehensive, and fees and tariffs are subject to change.

As an importer, you'll need to consider customs bonds and duties:

Customs bond

If you're importing more than $2,500 worth of goods to sell within the United States, you will have to post a bond (*http://bit.ly/customs_bonds*) "to ensure that all duties, taxes, and fees owed to the federal government will be paid." There are two types of bonds: single-entry bond and continuous-entry bond (also known as annual bond). The values of bonds are based on the value of the cargo plus customs duties. Single-entry bonds are applicable to one imported shipment only. A continuous-entry bond covering $50,000 (the lowest value possible) costs approximately $250 per year and is effective for any number of shipments for up to one year. Most importers bringing in more than two shipments per year find continuous-entry bonds more cost-effective; consult with a qualified customs broker for advice on your unique situation. In addition to providing a Customs bond, customs brokers will also provide customs clearance service to clear your shipment through US Customs.

Customs duties

Customs duties are assessed based on the type of goods being imported and the value of the shipment. There is an international system of codes called the *Harmonized System* (HS). Each product you import must have an HS code. The HS system (*http://hts.usitc.gov/*) has 22 sections and 99 chapters and is incredibly detailed. Searching "watch," for example, returns 443 results covering all manner of variants of watches: with automatic winding, having over 17 jewels in the movement, having over 1 jewel but not over 7 jewels in the movement, of textile material, of base metal, and so on. Every facet of what it means to be a watch, and every permutation of those facets, is assigned a unique HS code.

Each country sets its own tariff rates with reference to the HS system (in the US, this is called the Harmonized Tariff Schedule), which means different countries could have different tariff rates and Customs duties based on the value of the shipment. It is therefore necessary to be sure that the product you're importing is classified properly. Where the item is being imported from is also highly relevant; tariff rates for a particular HS code vary based on where it is made. The system is extremely complex and mistakes can be costly. If you are importing an innovative new type of device that doesn't fit neatly into a category, picking an HS code can be a daunting task. For new products that do not match with an existing classification, importers can obtain a unique binding ruling (*http://bit.ly/binding_ruling_filing*) from US Customs.

Other common importation fees include Merchandise Processing Fees (MPF) of .3464 percent of the shipment value (minimum $25, maximum $485) and Harbor Maintenance Fees (HMF) of .125 percent of the shipment value, applied to ocean freight only.

Finally, there are customs assists (*http://learn.flexport.com/customs-assists/*). "Customs assists are goods or services provided by a buyer that lend tangible value to the production of the final product by the manufacturer," according to Ryan Petersen, founder of freight forwarder and customs brokerage startup Flexport (*https://www.flexport.com/*). "If the U.S. company assisted in the design of the product, or provided any kind of tools or services at a discount to their market value, then those costs have to be added back to the value of the goods." These assists are critical to determining the *fair dutiable value* of the shipment. In the case of hardware, engineering or design services that happened outside of the US and were used in product manufacturing must be added into the value of the shipment (this is true of drawings for apparel production as well). Software development work is considered *intangible*, since it does not physically affect the product.

Ultimately, it's the importer's responsibility to get all of the papers filed and duties paid correctly. A mistake can result in goods being held at the dock, fines, audits, or increased scrutiny of future shipments (potentially causing delays or other headaches). For example, a late Importer Security Filing (ISF, also called a *10+2 filing*) can result in a $5,000–$10,000 fine. Because of the complexity, many importers hire a customs broker to take care of the paperwork and filings for them. "Depending on what product you are importing and from which country, there are 120 different forms that may need to be filed involving 10 government agencies," says Flexport's website. The site offers a thorough glossary (*http://learn.flexport.com/glossary/*) of terms that can help founders who are new to importing learn the lingo and understand the fees.

Regulatory Concerns and Certification

Since "hardware" is such a broad category, it's difficult to cover all of the possible potential regulatory concerns associated with devices in one place. If you're producing a health-related wearable or medical device, there's a good chance you'll be subject to some form of FDA regulation. Drone technology may be subject to FAA regulation (*https://www.faa.gov/uas/*).

Besides sector-specific regulation, industry certifications and compliance checks might also be required before you can sell your product (either domestically or overseas). We'll touch on some of the more common ones in this section. There are also many technical factors that you will need to consider while going through certification. Be sure to read "Certification" on page 134 for more context on what your engineers should concentrate on while going through certification.

MEDICAL DEVICES AND THE FDA

The FDA is a regulatory body responsible for protecting public health. It regulates everything from vaccines and pharmaceuticals, to medical processes such as blood transfusions, to medical devices themselves. The Center for Devices and Radiological Health (http://bit.ly/about_cdrh) is the subdivision of the FDA that monitors medical devices (ranging from toothbrushes to pacemakers), as well as radiation-emitting common household devices such as cell phones and microwaves. The definition of a medical device (http://bit.ly/fda_med_device) encompasses any product that is used to diagnose, cure, treat, mitigate, or prevent disease or other conditions in humans or animals, but does not achieve its purpose through chemical action or metabolic processes. The FDA classifies devices (http://bit.ly/med_device_class) into one of three categories, reflecting the level of potential risk associated with the product:

Class I
  The lowest risk, requiring the least regulation. Dental floss and adhesive bandages fall into this category.

Class II
  Slightly higher risk, and the FDA regulates to assure safety and effectiveness. Soft contact lenses are a Class II device.

Class III
  The highest risk. Misuse can cause serious harm, so the FDA typically requires that the device go through a formal evaluation. Pacemakers and stents are examples of a Class III device.

The FDA maintains a classification database (http://bit.ly/fda_classification) to help device manufacturers determine what class they are likely to be assigned.

If FDA review is needed, the device will be subject to one of two processes (http://bit.ly/fda_device_review). The first is *clearing*, in which the FDA reviews a document called a 510(k), also known as a *premarket notification*. The 510(k) submission demonstrates that the device is "substantially equivalent to a device that is already legally marketed for the same use." It must be filed a minimum of 90 days before marketing the product. Some categories of common, low-risk devices are 510(k) exempt (http://bit.ly/501k_exempt_device). This means that they don't need this submission to come to market, provided their packaging and labeling meets certain requirements, and they are manufactured to FDA standards. Even 510(k) exempt devices file FDA registration forms, and they might be subject to other FDA regulations around areas such as marketing claims. (See "Cellscope: A Case Study" on page 287, in which Amy Sheng discusses her experiences developing a medical device.)

Cellscope: A Case Study

Cellscope is on a mission to build a smartphone-enabled diagnostic toolkit that will improve remote diagnosis capabilities, allowing patients to get rapid answers to pressing health questions wherever they may be. Patients use Cellscope's smartphone attachments at home to gather diagnostic-quality data and transmit it to their doctors. The team's first product is the Cellscope Oto, a digital otoscope that produces high-quality visual imagery and video of the ear. Here, founder Amy Sheng walks us through the unique legal and regulatory challenges faced by a hardware startup in the medical space.

Cellscope was conceived during cofounder Erik Douglas's postdoctoral research in the mobile microscopy lab at UC Berkeley. The lab was interested in the potential of turning ordinary cellphones into microscopes and using them as portable diagnostic devices in poor and remote regions of the world. It was working on remote diagnosis of tuberculosis and malaria by visible inspection.

Amy and Erik believed that there was another large market for the smartphone-based technology: new parents, who must take children to the pediatrician's office for many childhood ailments. Besides the challenge of getting appointments and the hassle of travel, children with immature immune systems may catch other diseases while sitting in the waiting room. They envisioned a home toolkit that parents could reach for to transmit data to the doctor without having to leave the house, something that would help them determine whether an ailment was serious enough to warrant an office visit. After an evaluation of the market, they decided to build an otoscope, the tool that doctors use to do a visual examination of the inner ear as they diagnosis ear infections. They secured the right to license the technology from the Office of Technology Licensing at Berkeley and formed a company.

From the start, Cellscope's vision involved the empowerment of consumers. Amy remembers, "We wanted to give them tools that currently exist only in hospitals, or for doctors, and enable them to be more proactive about the health and their children's health." However, as the team progressed down the path of development, they quickly found significant regulatory challenges associated with bringing a diagnostic product directly to the consumer market.

They decided to run their pilot program with a group of pioneering physicians, who could use the device in their clinics and share feedback with the Cellscope team.

"We're very proud of our hardware," Amy says. "We had to solve some really tricky, big problems to make a device that can take diagnostic-quality images and video down a deep dark hole, which is what your ear is. But as a company, our vision is not to be a hardware medical device manufacturer." They view the hardware as a tool for data gathering and are most excited about the potential for clinical decision support software and other analytics tools. Data gathered by the Cellscope Oto has provided the team with what is, to their knowledge, one of the world's largest databases of ear images and videos. They are analyzing that data and using it to improve diagnostic capabilities.

The Cellscope software platform includes several components. There's a mobile app on the phone, which enables the user to capture imagery and video. Data gathered by the app gets transmitted to the web backend. The backend has two user categories: physicians and patients. Physician functionality includes the ability to review patient images. Consumers can see a library of their (or their child's) ears over time. The data analytics feature set is constantly being refined. The team is excited about the potential to use machine learning techniques to discover images that are similar.

Because of the sensitive nature of diagnostic devices, the Cellscope team began to work with an FDA regulatory consultant early on. The FDA classifies medical devices in a three-class system according to the likelihood that they could do harm. The amount of regulation and the quantity and type of documentation, submissions, and updates to the FDA vary depending on the class. Class 1 is the lowest-risk tier: it includes things like bedpans and Band-Aids... and video otoscopes. Even though the class indicates a low-risk device, it is occasionally necessary to submit what's known as a 510(k) to the FDA. This submission aims to prove that the device that will be marketed is safe and effective because it is sufficiently similar—"substantially equivalent" (SE) in FDA parlance—to another device already on the market that's already been approved. Video otoscopes are already classified as Class 1, and otoscopes are exempt (*http://bit.ly/501k_exempt_notes*) from the 510(k) process. Since Cellscope's Oto met the criteria for being an external ear canal otoscope, it didn't have to go through the 510(k) process.

Even though the Cellscope team didn't have to pursue 510(k) clearance, they are still subject to FDA regulatory guidelines. "Even if you're a Class 1 device, if you're a medical device company, you have to be registered with the

FDA," Amy says. "We have to follow what's called General Controls." General Controls is a list of things that a medical device company has to have in place. These include Standard Operating Procedures, which include frameworks for design control, as well as rules around collecting marketing inputs from the field and translating those inputs into requirements for both hardware and software. There are marketing requirement specifications (no outrageous claims) and specifications for both the hardware and software components of the device. There are verification and validation documentation requirements. "As a controlled medical device company, it's all about making sure that the end product is safe for consumers, for the general public, and for doctors," Amy says. "The process for putting documentation in place is so that you can go back and be able to trace where the problem started if something goes wrong."

The Cellscope team hired a regulatory consultant to help them understand the myriad rules that applied to their product. They also hired a full-time quality engineer, whose job entails making sure that the quality assurance system runs smoothly and training new employees in the documentation and process requirements unique to working at a medical appliance startup. "One of the things I would recommend is that for critical early hires, like the mechanical engineer and the quality engineer, bring on people who have done it [worked at a medical device company] before and they'll make sure that your whole company is doing it appropriately," Amy notes. While consultants provide valuable expertise, it's frequently reactive: the founder often has to know what question to ask the consultant. Hiring someone with expertise in medical devices brings that deep knowledge in-house, where it can be applied to all processes right from the start.

While starting a medical device company requires more rigorous processes than those at a traditional consumer-electronics startup, the reward of producing a device that has a profound impact on people's health outcomes motivates entrepreneurs to dive in anyway. While it may seem daunting at first, seasoned healthcare entrepreneurs like Amy assure new founders that they'll quickly adapt.

The second process is the *Premarket Approval* (PMA), which hopefully culminates in FDA approval (as opposed to clearing). In this case, the device isn't sufficiently equivalent and must use scientific evidence to demonstrate that it is both safe and effective. This is a much longer, more expensive process, generally applicable to high-risk medical devices.

The FDA not only regulates the hardware. It also regulates apps (http://bit.ly/fda_regulates_mmas) and the software used by connected devices. The clearing and approval processes depend greatly on case-by-case specifics, so we'll end this section by advising you to hire a professional. There are FDA consultants who can help you identify and navigate the specific requirements for your device.

HARDWARE AND THE FCC

The Federal Communications Commission (FCC) is a US government agency that regulates domestic and international communications via radio, TV, satellite, wire, and cable. As part of its mandate, the FCC protects against "radio and broadcast pollution" (http://fcc.gov) by regulating devices that produce electromagnetic waves.

The FCC classifies some products as *intentional radiators*: devices that deliberately broadcast radio energy as part of their core functionality, such as cell phones, Bluetooth devices, and radios. Intentional radiators require FCC certification. *Unintentional radiators*, such as TVs, digital cameras, and gaming systems, may emit radio energy as a side effect. They may be subject to FCC *verification* (a less rigorous process) or be exempt entirely. The FCC's Office of Engineering and Technology Equipment Authorization (http://bit.ly/fcc_oetea) site has information about the distinction, as well as lists of facilities that can perform the necessary certification or other tests.

The FCC approval process may take anywhere from a few weeks to a few months and can cost tens of thousands of dollars. Don't make the mistake of thinking about—or budgeting for—this process at the last minute.

Hardware startups face a daunting array of legal, regulatory, and tax-related issues that software startups typically don't encounter. Don't put off thinking about the issues discussed here until the last minute. There are so many complexities involved in building a device as it is. You don't want to be hit with unforeseen delays and costs due to regulatory or import/export errors. The main takeaway from this chapter should be that it's especially important to build out a support team of experienced lawyers who can help you navigate these challenges on your journey to market.

Epilogue: The Third Industrial Revolution

Globalization and economic changes have helped to turn the maker movement into something profoundly important on a *global* scale. Between 1760 and 1830, the First Industrial Revolution introduced mechanized manufacturing and sparked a transition away from human-powered production. Whether we assign the Second Industrial Revolution to Henry Ford's introduction of the modern assembly line and mass production or to the democratization of the personal computer and the beginnings of the Internet (also called the "Digital Revolution"), all three of these periods in time have had profound economic implications, particularly with respect to jobs.

The Third Industrial Revolution (*http://bit.ly/3rd_industrial_rev*) is about the power of the Web intersecting with modern manufacturing technologies. Digital manufacturing has serious economic impacts. It has shifted the geography of production and changed the face of the modern factory into one that requires fewer workers. Workers who operate the factories of the future will require a different kind of technical skill. They will be monitoring and operating the robots and machines that make the products, not physically working the line themselves.

Political leaders are paying attention. President Obama referenced the future of manufacturing in his 2013 State of the Union address, announcing an initiative to create new manufacturing hubs and high-tech jobs within the US. While factory jobs will change, the advances of the Third Industrial Revolution have opened the door for new types of companies to emerge.

Index

Symbols
3D printing, 109
3D Robotics, 14, 31-32
4 Ps of Marketing, 229
500 Startups, 139
7UP, 74

A
Aaker, Jennifer, 65
acceleration/accelerators, 7, 139-154
    about, 139
    AlphaLab Gear, 145
    choosing incubator/accelerator, 150-154
    Flextronics, 149
    HAXLR8R, 144
    Highway1, 147
    incubators vs., 139
    Lemnos Labs, 142
    littleBits case study, 141-142
    PCH, 146-149
    PCH Access, 148
accelerometer, 112
accessible markets, 43
accessories, supplementary, 215
activation fee, 217
Adafruit, 5
additive manufacturing, 109
addressable market, total market vs., 205
advertising, 230
aggregator marketplaces, 245-248
agricultural robots, 14
Agrobot, 14
air freight, 129
Airware, 143
Alibaba, 6, 129
AlphaLab Gear, 145
Altium, 84
Amazon Flexible Payments, 160
Amazon Fulfillment, 265
Amazon Marketplace, 262
Amazon.com
    and Kindle Fire business model, 215
    price matching, 257
    reviews as customer research, 228
    selling through, 247
analysis paralysis, 86
analysis, of market, 39
analytics, 183
anchoring, 222
Anderson, Chris, 30-32, 217
Andon, Alex, 262-264
Android tablets, 215
angel investors, 196-200
angel syndicates, 196
AngelList, 197-200, 202
app-store model, 218
Apple, 2
    and iPad business model, 215
    and multi-channel pricing, 257
    iPad NPS, 244
    mission statement, 59
    packaging standards set by, 137
    product naming by, 67
    value of brand name, 54
approved vendor list (AVL), 115
archetype study, 65
Arduino, 5, 31, 89, 95, 103, 176
ARM, 96

Armour39, 12
Artoo, 103
Ashton, Kevin, 9
ASIC (application-specific integrated circuit) electrical component, 110
assembly line, 291
asset tracking, 10
assets, brand, 67-73
audience, identifying, 233
August, 10
authenticity, 64
Automatic, 10
autonomous robots, 13
AVL (approved vendor list), 115

B

B2B (business-to-business) model, 32
    sales cycles and profit margins, 224
    value-based pricing and, 226
B2C (business-to-customer) model, 227
backend, 102
BackerKit, 186
Baker, Phil, 275
Bar Test, 69
Barros, Marc, 65, 69-72, 237, 261
Baxter robot, 13
Be Like Mike vs. Be Like Joe strategy, 237
BeagleBone, 5, 89
BeagleBone Black, 95
behavioral profiles, 227
behavioral segmentation, 45
Belkin, 10
Best Buy, 76, 253, 259
Beta List, 50
Betabrand, 242
Bezos, Jeff, 215
BGA (ball grid array) electrical component, 110
big-box retailers, 249-261
    and startups, 76
    Nest case study, 252
    Slam Brands case study, 250
bill of materials (BOM), 117
Blank, Steve, 17, 39, 51
Bluetooth, 100
bootstrapping, 191

boutique manufacturing, 124-128
brainstorming, 68
brand assets, 67-73
brand differentiation, 74-77
brand equity, 54
brand experience, 74
brand ideal, 60
brand identity, 61-65, 243
brand image, 61
brand loyalty, 55
brand mantra, 58
brand personality, 65-67
brand positioning, 74-77
Brand Touchpoint Matrix, 73
brand touchpoints, 73-74
branding, 53-77
    and company's mission, 58-61
    basics, 53-58
    brand assets/touchpoints, 67-74
    brand identity/personality, 61-67
    Countour and Moment case study, 69-72
    J. Walter Thompson case study, 62-65
    naming your company, 67-72
    Nest case study, 75-77
    positioning/differentiation, 74-77
Branson, Richard, 60
breadboard, 85, 111
broad-spectrum accelerators, 139
budget, marketing, 238
build-measure-learn feedback loop, 17
Burch, Tory, 216
burn rate, 169
business models for hardware startups, 214-220
    device as platform for upselling other devices, 216
    OSHW, 219
    selling additional physical products, 215-217
    selling data, 219
    selling device plus ancillary materials, 216
    selling device plus digital content, 218
    selling device plus subscription services, 217

selling device plus supplementary accessories, 215
selling device to consumers and data to third parties, 219
selling only data to customers, 219
selling services or content, 217
buyers market, 231

C

C-Corporation, 270
CAC (customer acquisition cost), 43
CAD (computer-aided design), 5, 84, 115
campaign page, crowdfunding
   driving traffic to, 175-183
   marketing materials on, 172-175
cap, 211
capacitive switch, 112
Casey, Liam, 262
casting, 109
CBA (MIT Center for Bits and Atoms), 3
CDMA (code division multiple access), 101
CE (European Conformity), 135
Celery, 179
Cellscope, 287-289
cellular modem, 101
certification, 134-136
champions, 165
change orders (COs), 118
channel selection, marketing, 233-236
   Facebook, 234
   Google AdWords, 234-236
   Twitter, 236
Charlton, Erik, 75
Chibitronics, 124
China
   and Jellyfish Art case study, 263
   boutique manufacturing case study, 124-128
   manufacturing in, 115, 120-128, 178
Chumby, 124
Cisco Systems, 9
CIT, 260
clearing, 286
Clerky, 273

click-through rate (CTR), 236
cloud-based fabbing services, 5
Clover, 85
CM (contract manufacturer), 114
CNC (Computer Numerical Control), 109
Coca-Cola, 54
cofounders, 21-23
COGS (cost of goods sold), 117, 169, 223
Comcast, 217
commissions, rep group, 256
commitment, of cofounders, 23
communities
   and user retention, 231
   computer technology (1970s-1980s), 2
   online, 7
   updates when crowdfunding, 186
community building, 187
community engagement
   3D Robotics case study, 31-32
   and cofounders, 21-23
   early, 26-33
   idea validation and, 17-33
   OpenROV case study, 27-29
   with mentors, 23-26
   with other hardware founders, 20
community manager, 64
company, forming as legal entity, 270-274
competition, 206
components
   electrical, 110
   for prototyping, 92-93
   inexpensive, 5
compression molding, 109
computer-aided design (CAD), 5, 84, 115
connected devices, 9-11
connectivity
   connected devices, 9-11
   prototyping for, 98-102
Conrad, Jeremy, 142
consolidators, 130
content plays, 218
Contour Cameras, 65
   marketing message, 237
   naming case study, 69

contract manufacturer (CM), 114
conversation, validated learning and, 20
convertible debt, 210
convertible note ask, 210
convertible note, priced equity round vs., 211
copywriting, crowdfunding, 173
Corrado, Ben, 92, 135
COs (change orders), 118
cost of goods sold (COGS), 117, 169, 223
cost-plus pricing, 223
Craigslist, 50
critical-to-function (CTF), 132
Croll, Alistair, 51
Cross, Sean "xobs", 126
Crossing the Chasm (Moore), 75
crowdfunding, 155-188
    and email lists, 175
    and news media, 179-183
    and social media, 175
    campaign page marketing materials, 172-175
    data-driven, 183-186
    DIY approach, 158-161
    donation-based vs. equity, 155
    driving traffic to campaign page, 175-183
    ecosystem for, 155-161
    financial model for, 169-171
    fundraising vs., 187
    Indiegogo, 157, 161-163
    Kickstarter, 156
    manufacturing and, 171
    Misfit Wearables case study, 165-166
    organizing PR materials, 180-181
    Pebble watch case study, 176-178
    perks, choosing, 161-164
    perks, pricing, 164, 167-169
    planning your campaign, 161-172
    publicizing your campaign, 175-183
    real value of platform, 187
    real-time adaptation, 183-186
    Scout Alarm case study, 158-160
    syncing with manufacturing, 171
    tips/tricks when using Indiegogo, 161-163
    understanding backers, 161-164
    updates for your community, 186
    while your campaign is live, 183-187
Crowdhoster, 158
CrunchBase, 202
CTF (critical-to-function), 132
CTR (click-through rate), 236
customer acquisition cost (CAC), 43
customer development, 46-51
Customer Development methodology, 17
customer research, 227
customer service, 266
customer, call to action for, 237
customer-focused market, 231
customs bonds/duties, 283-285
Cylon.js, 103

D

Dash Navigation, 244, 248, 259
dashboards, 183
data plays, 219
data, selling to third parties, 219
debt, funding via, 191
defensibility, 214
demand, crowdfunding for building, 187
demographic market segmentation, 44, 233
design for manufacture, 90
design for X (DFX), 114, 117
design prototype, 19
design verification test (DVT), 114, 119
designed products, 14
Designing Brand Identity (Wheeler), 54, 61
detailers, 259
DFX (design for X), 114, 117
DHL, 266
Dictionary of Brand (Neumeier), 61, 65
die cutting, 109
differentiation, brand, 74-77
differentiators, 41
digiKey, 92
digital content, selling, 218
digital manufacturing, 291
Digital Revolution, 291

Dimatos, John, 173, 182, 184, 187
direct sales, online, 241-244
disposables, 128
distribution channels, 8, 241-266
　aggregator marketplaces, 245-248
　big-box retail, 249-261
　Grand St. case study, 245
　Jellyfish Art fulfillment case study, 262-266
　Nest case study, 252
　online direct sales, 241-244
　online specialty retailers, 245-248
Dixon, Chris, 40, 215
DIY crowdfunding, 158-161
DIY Drones, 30
do-it-yourself (DIY), 2
Docracy, 273
donation-based crowdfunding, 155
Dougherty, Dale, 3, 8
Dragon Certified, 172
Dragon Innovation, 160, 172, 178
Drane, Kate, 161
drill files, 91
drip campaign, 242
DroneDeploy, 14
drones (unmanned aerial vehicles), 14, 31, 143
Dropcam, 107
dual licensing, 220
due diligence, 208
DVT (design verification test), 114, 119

E

e-tailers (online retailers), 245-248
EAGLE, 84
earned media, 229
Eat My Words, 68
Ebersweiler, Cyril, 144
Economic Development Councils (EDC), 195
ECOs (engineering change orders), 118
ecosystem, supplemental, 7
EDA (electronic design automation), 84
efficient entrepreneurship, 8
EIR (entrepreneur in residence), 21
Electric Imp, 5

electrical components, 110
electrical engineering (EE) prototyping, 89-90
electromagnetic interference (EMI), 135
electronic design automation (EDA), 84
electronic manufacturing service (EMS), 114
elevator pitch, 58
Ellsworth, Adam, 162
email, for crowdfunding, 175, 180
endcap, 253
endmills, 109
engineering change orders (ECOs), 118
engineering verification test (EVT), 114, 119
entrepreneur in residence (EIR), 21
entrepreneurship, efficient, 8
environmental sustainability, 133
equity crowdfunding, 155, 197
equity financing, 210
equity round, structuring, 210-212
equity, mentor, 25
Etsy, 8
Evans, Nick, 239
EVT (engineering verification test), 114, 119
exporting, 283-285
extrusion, 109

F

FAA regulation, 285
fab labs, 3, 21
Facebook, 50
　and crowdfunding updates, 175
　and Tile marketing campaign, 240
　for True Believers, 30
　in-stream advertising, 234
factor financing (factoring), 260
Fadell, Tony, 76
FAI (first article inspection), 132
FAIR (first article inspection report), 132
fair dutiable value, 285
family, as source of funding, 195
Farley, Mike, 239

FCC verification, 290
FDA (Food and Drug Administration), 136, 286-290
FDA device classification, 286
feature comparison matrix, 41
Federal Communications Commission (FCC), 134, 290
FedEx, 266
fees, crowdfunding, 170
Feld, Brad, 210, 214
Fenwick & West, 273
Fetch Robotics, 13
financial model, crowdfunding, 169-171
firmware (FW), 90, 102, 108
first article inspection (FAI), 132
first article inspection report (FAIR), 132
First Industrial Revolution, 291
first-to-file approach, 274
Fisher Law Offices, 270
Fisher, Josh, 270, 275
Fitbit, 216, 259
five-factor model, 65
fixed-funding campaign, 157
fixtures, tooling, 131
flexible campaign, 157
Flexport, 285
Flextronics, 149
flying robots, 14
FOB (free on board), 130
Food and Drug Administration (FDA), 136, 286-290
Ford, Henry, 291
foreign manufacturers, 129
Formlabs, 3
Forrest, Brady, 22, 147
Four Ps (crowdfunding participation), 162
Four Steps to the Epiphany (Blank), 51
free on board (FOB), 130
Free: The Future of a Radical Price (Anderson), 217
FreeRTOS, 95, 95
Freescale, 96
freight forwarders, 130
friends, as source of funding, 195
From Concept to Consumer (Baker), 275

Fry's, 248
FuelBand, 12
Fulfilled by Amazon, 247, 262, 265
funding portals, 197
fundraising, 189-212
  angel investors, 196-200
  AngelList, 197-200
  bootstrapping, 191
  crowdfunding vs., 187
  debt, 191
  due diligence process, 208
  from friends/family, 195
  grants, 191-195
  JOBS Act requirements, 197
  personalized introductions to venture capitalists, 202-204
  platforms, 7
  SBIR program, 191-195
  story, telling your, 204-208
  strategic investors, 209
  structuring your round, 210-212
  targeting venture capital investors, 201
  timing and priorities, 190
  venture capital, 201-209
FW (firmware), 90, 102, 108

G

Gallo, Carmine, 60
gateways, 102
General Motors, 13
Gerber files, 91
Gershenfeld, Neil, 3
gestural prototype, 84
Gillette, 216
Global Network Navigator, 3
Gobot, 103
going to market, 213-267
  aggregator marketplaces, 245-248
  audience, identifying, 233
  big-box retail, 249-261
  business models for hardware start-ups, 214-220
  channel selection, 233-236
  defining your objective, 231
  distribution channels and related marketing strategies, 241-266
  Grand St. case study, 245

Jellyfish Art case study, 262-264
KPI (key performance indicator) choices, 232
marketing, 228-241
message formulation, 237
Nest case study, 252
online direct sales, 241-244
online specialty retailers, 245-248
OSHW business model, 219
pricing, 220-228
refining your campaign, 238
rep groups, 254-258
Slam Brands case study, 250
small retailers and specialty shops, 248
Tile case study, 239-241
timeline, 238
warehousing and fulfillment, 261-266
golden samples, 132
Goldwater, Dan, 120-122
Gongkai, 126
Goodwin Proctor, 274
Google, 10
  mission statement, 59
  value of brand name, 54
Google AdWords, 50, 234-236, 240
Google Analytics, 50, 239
Google Groups, 30
Google+, 175
GoPro, 70, 237
GPS (global positioning system), 101, 112
Grand St., 8, 245
grants, 191-195
graphic designer, 137
greenwashing, 134
gross margin, 169
Grow (Stengel), 60
growth-stage companies, 212
GSM (Global System for Mobile Communications), 101
Gunderson Dettmer, 274
gyroscope, 112

H

hackathons, 22
hacker for hire, 21
Hackerspaces, 7
HackerspaceWiki, 21
Hacks (book series), 3
Harbor Maintenance Fees (HMF), 284
hardware
  FCC regulation of, 290
  open-source, 6
hardwarians, 20
Harmonized System (HS), 284
Haven, 130
HaxAsia, 150
HAXLR8R
  accelerator, 144
Hello Future, 73
Hermès, 221
Hero and the Outlaw (Mark and Pearson), 65
Highway1, 147
HMF (Harbor Maintenance Fees), 284
Homebrew Computer Club, 2
HS (Harmonized System), 284
Hsieh, Tony, 60
Huang, Andrew "bunnie", 118, 124-128
Huffman, Todd, 192
Hunch, 40

I

IAO variables, 44
iBeacon, 102
ICs (integrated circuits), 94-98
ID (industrial design), 87
idea validation
  and community engagement, 17-33
  and Lean Startup, 17-20
  cofounders, 21-23
  from other hardware founders, 20
  from True Believers/early community, 26-33
  mentors and, 23-26
  OpenROV case study, 27-29
  teams, 21-23
identifying your audience, 233
IDEO, 80-82
iFixit, 86
importing, 129, 283-285
inbound marketing, 229
incoming quality control, 133

incubators, 139
    accelerators vs., 139
    choosing, 150-154
Indiegogo, 157
    dashboards, 183
    fees, 170
    fundraising goal recommendations, 183
    Misfit Wearables and, 166
    pledge levels, 164
    tips/tricks when using, 161-163
    video guidance on, 173, 175
Indiegogo Outpost, 158
Indiegogo Playbook, 157
industrial design (ID), 87
Industrial Internet, 10
Industrial Revolution, 291
inexpensive components, 5
ingress protection (IP), 117
innovation divisions, retailer, 258
insourcing, for prototyping, 90
Instagram, 175
Instructables, 3
integrated circuits (ICs), 94-98
Integrity, 95
intellectual property (IP)
    and NDAs, 120
    Western vs. Gongkai approach to, 126
Intelligent Products Marketing (IPM), 254
intentional radiators, 290
interaction design (IxD), 88
interfaces, 102
intermodal freight transport, 129
international market, 166, 181
Internet of Things, 5, 9
introductions, to venture capitalists, 202-204
investors, online sales metrics for, 243
IP (ingress protection), 117
iPad, 215
    and sellers market, 230
    NPS for, 244
IPM (Intelligent Products Marketing), 254
Iris Smart Home Management System, 10, 217
iRobot, 191

iterative process
    EMI shielding, 135
    Lean Startup development as, 18
    tooling as, 131

J

J. Walter Thompson, 62-65
Jawbone, 42, 58
Jellyfish Art fulfillment case study, 262-264
jigs, 131
JOBS Act, 197
Jordan, Michael, 237
junior investors, 203

K

Kahneman, Daniel, 222
Kaplan EdTech Accelerator, 140
Kawasaki, Guy, 207
Kelly, Kevin, 11
key performance indicators (KPI), 232, 238
Keyword Planner, 235
keyword(s), 234
Kickstarter, 156
    and Raspberry Pi, 92
    dashboards, 183
    fees, 170
    Jellyfish Art and, 263
    Otherfab and, 168
    Pebble watch and, 176-178
    pledge levels, 164
    video guidance on, 173, 175
Kickstarter Status Board, 186
Kicktraq, 186
Kindle Fire, 215
KISSmetrics, 239, 243
Klein, Eric, 244, 248, 259
knockoffs, 214
KPI (key performance indicators), 232, 238

L

Lab IX (Flextronics acceleration platform), 149
landscape for hardware startups, 1-15
    connected devices, 9-11

current hardware companies, 9-15
designed products, 14
efficient entrepreneurship, 8
history/background, 1-4
inexpensive components, 5
lean startup, 8
maker movement, 1-4
open-source hardware, 6
rapid prototyping, 5
robotics, 13-14
small-batch manufacturing, 6
supplemental ecosystem, 7
technology-assisted scale, 4-8
wearables, 11
Lang, David, 26, 27-29
large retailers, 243
laser cutting, 110
Launch Rock, 50
lead angels, 198
Lean Analytics (Croll and Yoskovitz), 51
lean manufacturing, 19
Lean Startup, 8, 17-20, 39
Leap Motion, 103
legal issues, 269-290
    Cellscope regulatory case study, 287-289
    company formation, 270-274
    FCC regulations, 290
    import/export considerations, 283-285
    liability, 281
    manufacturing concerns, 281-285
    medical devices and FDA, 286-290
    patents, 275-281
    regulatory concerns and certification, 285-290
    trade secrets, 275
    trademarks, 274
LegalZoom, 273
Lemelson, Jason, 250-251, 258
Lemnos Labs, 142
letters of intent (LOI), 206
Levine, Steven, 254-256
liability, 281
life cycle, product, 128
life hacking, 2
lifetime value (LTV), 43
LightBlue Bean, 100

limit switch, 112
limited liability, 270
limited partners (LPs), 201
LinkedIn, 23, 50
littleBits, 141-142
LLC (limited liability company), 271
location, for manufacturing, 120-128
lock-in, 55
Lockitron, 158
Logistica Asia, 150
LOI (letters of intent), 206
looks-like prototypes, 85
loss aversion, 222
Lowenstein Sandler, 274
Lowes, 217, 258
LPs (limited partners), 201
LTV (lifetime value), 43
Lumo BodyTech, 47-50, 216
Lyons, Dave, 85, 123

M

machine-to-machine (M2M) devices, 9
magnetometer, 112
Make magazine, 3
Maker Faire, 4
Maker Map, 21
maker movement
    and Whole Earth Catalog, 2
    early computer communities, 2
    globalization of, 291
    Make magazine, 3
    MIT Center for Bits and Atoms, 3
MakerBot, 5
makerspaces, 7
mantra, 58
manual insertion, 111
manufacturing, 113
    agreements and contracts, 282-283
    and rapid prototyping, 5
    boutique projects in China case study, 124-126
    certification, 134-136
    choosing location for, 120-128
    import/export considerations, 283-285
    importing from foreign manufacturers, 129
    legal issues, 281-285

302 | INDEX

liability issues, 281
MonkeyLectric case study, 121-122
Orion Labs case study, 115
packaging, 136-138
preparations for, 114-120
prototyping vs., 113
small-batch, 6
supply chain management, 128-129
sustaining, 138
syncing crowdfunding timeline with, 171
technical issues to address during, 131-134
terms about processes, 108-110
manufacturing agreements, 282
margin, 224
Mark, Margaret, 65
market
   addressable, 205
   and venture capital pitch, 205
   behavioral segmentation, 45
   customer acquisition cost, 43
   customer development, 46-51
   demographic/psychographics, 44
   differentiators, 41
   knowing your, 35-51
   lifetime value, 43
   LumoBodyTech case study, 47-50
   market analysis, 39
   questions for, 36
   researching, 36-43
   segmenting your market, 43-46
   size, 36-37
   trajectory, 38
   who/what/why of your product, 36
market analysis, 39
market development funds (MDF), 253, 259
market requirements document (MRD), 118
market research, 36-43
   crowdfunding as, 187
   differentiators, 41
   for value-based pricing, 226-228
   market analysis, 39
   market size, 36-37
   market trajectory, 38
market segmentation, 43-46
   behavioral segmentation, 45

customer acquisition cost/lifetime value, 43
defined, 43
demographics/psychographics, 44
market size, 36-37
market trajectory, 36, 38
market-based pricing, 225
marketing, 228-241
   and compliance, 134
   audience, identifying, 233
   budget, 238
   call to action, 237
   channel selection, 233-236
   crowdfunding campaign page, 172-175
   defined, 229
   defining your objective, 231
   KPI choices, 232
   message formulation, 237
   refining your campaign, 238
   Tile case study, 239-241
   timeline, 238
   with Facebook, 234
   with Google AdWords, 234-236
   with Twitter, 236
Markoff, John, 2
Marks, James, 265
Mason, Chris, 254-257
mass production, 291
master carton, 137
Maurya, Ash, 36
McCarthy, E. Jerome, 229
McMaster-Carr, 92
MCro, 12
MDF (market development funds), 253, 259
mechanical engineering (ME) prototyping, 89-90
Mechanical Turk, 50
media (news), for crowdfunding publicity, 179-180
medical devices, 286-290
Meetup, 50
Mendelson, Jason, 210
mentor(s), 23-26
Merchandise Processing Fees (MPF), 284
metrics, online sales, 243
Mexico, manufacturing in, 120

microcontroller, 95
microprocessor, 95
MicroPython, 96
Microsoft, 59
Migicovsky, Eric, 151, 176-178
milling, 109
minimum viable product (MVP), 18
misappropriation, 275
Misfit Wearables
    analytics used by, 185
    and supplementary accessories, 216
    crowdfunding case study, 165-166
    customer research, 228
    international sales, 181
mission of company, 58-61
mission statement, 58
MIT App Inventor, 83
MIT Center for Bits and Atoms (CBA), 3
ModCloth, 242
molding, 109
Moment (brand), 71
Moore, Geoffrey, 75
Morgan Stanley, 9
Mouser, 92
MPF (Merchandise Processing Fees), 284
MRD (market requirements document), 118
multimeter, 94
Muñoz, Jordi, 31
Murphy, Sean, 55-58, 59
MVP (minimum viable product), 18

N

naming your company, 67-72
NDAs (non-disclosure agreements), 120
Near Field Communication (NFC), 101
Nest, 10, 107
    branding case study, 75-77
    distribution channels case study, 252
net 30 terms, 252
Net Promoter Score (NPS), 243
Neumeier, Marty, 61, 65
new product introductions (NPIs), 88

news media, for crowdfunding publicity, 179-180
Next Level Sales and Marketing, 254
NFC (Near Field Communication), 101
Nicholson, Chris, 179-180
Nielsen, 45
Nike, 59
Nike+, 12
Ning, 28, 30
Nivi, Babk, 197
Njoo, Enzo, 182
non-disclosure agreements (NDAs), 120
NPIs (new product introductions), 88
NPS (Net Promoter Score), 243
Nucleus, 95

O

Obama, Barack, 291
ocean freight, 129
ODM (original design manufacturer), 113
OEM (original equipment manufacturer), 113
off-the-shelf (OTS) solutions, 92
offline metrics, 243
OMS (order management system), 262
online community, 7, 20
online metrics, 243
online retailers, 245-248
OOBE (out-of-box experience), 137
open source hardware (OSHW), 6, 219
Open Source Hardware Association (OSHWA), 6
OpenROV, 14, 27-29
operations team, for sustaining manufacturing, 138
order management system (OMS), 262
original design manufacturer (ODM), 113
original equipment manufacturer (OEM), 113
Orion Labs, 115
Orrick, 25, 273
oscilloscope, 94
OSHW (open source hardware), 6, 219
OSHWA (Open Source Hardware Association), 6

Otherfab, 168
Otherlab, 3
OTS (off-the-shelf) solutions, 92
out-of-box experience (OOBE), 137
outbound marketing, 230
outsourcing, for prototyping, 90
Ouya, 218
owned media, 229

P

packaging, 136-138
packaging designer, 137
paid media, 230
pallets, 137
Pampers, 61
pass-through companies, 271
pass-through taxation, 271
patents, 275-281
pay-per-click advertising platform, 235
PCB (printed circuit board), 88, 111
PCBA (printed circuit board assembly), 111, 132
PCH, 146-149
PCH Access, 115, 148
Pearson, Carol, 65
Pebble, 40, 103
Pebble watch, 95
 crowdfunding case study, 176-178
 international sales, 181
Pendergast, David, 274, 282
perks, crowdfunding
 choosing, 161-164
 pricing, 164, 167-169
persona building, 45
personal sensor devices (wearables), 11
Petersen, Ryan, 285
Peyton, Amanda, 245
Phillips, Robert, 226
PhoneGap, 83, 103
Photon, 97
Pinoccio, 171
Pinterest, 175
pitches, 204
platform play, 216
pledge levels, 164
PMA (Premarket Approval), 289
Polaroid, 275
Ponoko, 5

positioning, brand, 74-77, 221
PRD (product requirements document), 118
Premarket Approval (PMA), 289
premarket notification, 286
press releases, for crowdfunding PR, 181
price optimization, 222
price-matching, 257
priced equity round
 convertible note vs., 211
 structuring your round, 210-212
pricing, 220-228
 cost-plus, 223
 crowdfunding perks, 164, 167-169
 market-based, 225
 of crowdfunding reward tiers, 169-171
 value-based, 226-228
Pricing and Revenue Optimization (Phillips), 226
Pricing Strategy (Smith), 220
printed circuit board (PCB), 88, 111
printed circuit board assembly (PCBA), 111, 132
printed circuit boards (PCBs), 89
prioritized examination option, 280
privacy
 and data plays, 219
 software, 107
PRIZM, 45
proactive community-building, 187
product requirements document (PRD), 118
product, project vs., 4
product-focused market, 230
product/market fit, 36
production verification test (PVT), 114, 119
profit margin, 224
project, product vs., 4
proof-of-concept, 84
prototyping, 79-112
 and connectivity, 98-102
 and mechanical/electrical engineering, 89-90
 common terms used in, 84
 design prototype, 19

electrical component terminology, 110
glossary of terms, 108-112
IDEO case study, 80-82
industrial design, 87
integrated circuits for, 94-98
manufacturing vs., 113-138
outsourcing vs. insourcing, 90
parts/components for, 92-93
process terminology, 108-110
rapid, 5
reasons for, 79-83
sensor terminology, 111
software platforms, 102-107
software security/privacy, 107
software team for, 90
Spark case study, 97
team for, 87-90
teardowns, 86
tools for, 93
types of, 83-87
UX/UI/IxD, 88
works-like/looks-like prototypes, 85
psychographics, 44, 227
psychological research, 65
publicity, for crowdfunding campaign, 175-183
Punch Through Design, 100
PVT (production verification test), 114, 119

Q

QFN (quad flat no-leads) electrical component, 111
Qi, Jie, 125
Qualcomm, 191
qualitative research, 56
quality (term), 133
quality assurance (QA), 133
quality control (QC), 133
Quantified Self movement, 11
quantitative research, 57
Quirky, 14
Quora, 23, 39, 202

R

radio function (RF), 96
rapid prototyping, 5

Raspberry Pi, 5, 89, 92, 95
Ravikant, Naval, 197
razors-and-blades lock-in model, 216
reactive community-building, 187
real-time operating system (RTOS), 95
Reddit, 50
Reebok, 12
regulation(s), 285-290
    FCC, 290
    medical devices and FDA, 286-290
remotely operated vehicles (ROV), 14
rent, 262
rep groups, 250, 254-258
repairs, 128
request for quote (RFQ), 117
research, branding, 62-64
research, product
    qualitative, 56
    quantitative, 57
researching, marketing, 36-43
retail aggregator platforms, 245-248
retail sales, 241-266
    aggregator marketplaces, 245-248
    big-box retail, 249-261
    Grand St. case study, 245
    Nest case study, 252
    online direct sales, 241-244
    rep groups, 254-258
    Slam Brands case study, 250
    small retailers and specialty shops, 248
    warehousing and fulfillment, 261-266
Rethink Robotics, 13
return on investment (ROI), 227
revenue, 206
rework, 111
RF (radio function), 96
RFQ (request for quote), 117
Ries, Eric, 8, 17
Ringelmann, Danae, 161-163, 175, 187
Roberts, Daniel, 158-160
Roberts, Ryan, 271
robotics, 13-14, 31-32
Rock Health, 140
Rogers, Matt, 75, 252
ROI (return on investment), 227
Romo, 9
Roomba, 103

rotary switch, 112
ROV (remotely operated vehicles), 14
RTD (resistance temperature detectors), 112
RTOS (real-time operating system), 95
ruggedization, 117
Running Lean (Maurya), 36

S

S-Corporation, 271
sales cycles, 224
sales volume, 225
SAM (serviceable available market), 37
Samsung, 40, 99
Sandbox Industries, 159
Satmetrix, 244
SBA (Small Business Association), 191
scale
    and rapid prototyping, 5
    inexpensive components, 5
    open-source hardware, 6
    small-batch manufacturing, 6
    technology and, 4-8
scale models, 85
Schox, Jeffrey, 277
SCM (supply chain management), 128-129
Scout Alarm, 158-160
screen renderings, 88
Second Industrial Revolution, 291
security, software, 107
self-syndicates, 198
Selfstarter, 158, 239
sell-in relationship, 247
sell-through relationship, 247
seller's market, 230
sensitivity analysis, 39
sensors, 111
serviceable available market (SAM), 37
serviceable obtainable market (SOM), 37
Shal, Jinal, 62-65, 67
Shapeways, 5
Shapiro, Dave, 158-160
shareholder value, branding and, 53
Shenzen, 6
Sherman, Andy, 115
shielding, EMF, 135

shipping rates, 266
Shopify, 232, 262
ShopLocket, 8
SIC (standard industrial classification), 45
Sidekick, 186
Sifteo, 218
Simpson, Star, 115
size, of market, 36-37
Slam Brands, 250
Small Business Association (SBA), 191
Small Business Innovation Research (SBIR), 191
    fundraising with, 191-195
    navigating the process, 193-195
small retailers, 248
small-batch manufacturing, 6
SMART (Specific, Measurable, Actionable, Relevant, Time-bounded), 232
smart cars, 10
smart devices, 5
smart homes, 10
smart watches, 40
smartphones, 5
SmartThings, 10, 99
SMD (surface-mount device) electrical component, 111
SMILE & SCRATCH test, 68
Smith, Tim, 220
SMT (surface mount technology), 94, 111
social media, 165, 175
soft tooling, 131
software
    and defensibility of hardware, 214
    and lock-in, 55
    platforms for prototyping, 102-107
    security/privacy and prototyping, 107
software engineers, 22
software team, for prototyping, 90
software-focused incubator, 151
SOM (serviceable obtainable market), 37
Sony, 40
SOP (standard operating procedure), 115
Spark (prototyping case study), 97
Spark Core, 5, 97

Spark Photon, 95
SparkFun, 5
specialty retailers, online, 245-248
Springboard, 150
Stackpole, Eric, 27
stamping, 109
standard industrial classification (SIC), 45
standard operating procedure (SOP), 115
Startup Company Lawyer, 273
Startup Lawyer, 273
Stengel 50, 61
Stengel, Jim, 60
Storefront.com, 249
story, telling your, 204-208
storyboards, 88
strategic investors (strategics), 209
strategic pricing, 222
Stutzenberg, Eric, 108
subscription services, 217
supplemental ecosystem, 7
supplementary accessories, 215
supply chain, 121-122
supply chain management (SCM), 128-129
surface mount technology (SMT), 94, 111
sustainability, environmental, 133
sustaining engineering, 120
Suster, Mark, 203
Swarovski, 216
switch, 112
Symantec, 191

T

T0 parts, 131
TAM (total addressable market), 37
Taobao, 6
Target, 251, 258
teams
  and venture capital pitch, 204
  in early phase of startup, 21-23
teardowns, 86
technical partner/cofounder, 22
technical team, manufacturing issues for, 131-134
technology
  inexpensive components, 5
  open-source hardware, 6
  scale and, 4-8
  small-batch manufacturing, 6
technology-assisted scale
  online community, 7
  rapid prototyping, 5
  small-batch manufacturing, 6
TechShop, 7, 21, 82
temperature sensor, 112
temporary switch, 112
Tesla Motors, 123
Tessel, 96
The Hybrid Group, 103
thermostats, 75-77
ThingMagic, 3
Third Industrial Revolution, 291
third parties, selling data to, 219
Three I's for publicizing startups, 179
through-hole electrical component, 111
TI, 96
Tile (marketing case study), 239-241
tilt switch, 112
timeline, marketing, 238
timing of fundraising, 190
Tindie, 8
toggle, 112
tool neutral design, 132
tool(s), prototyping, 93
tooling, 131
total addressable market (TAM), 37
total available market, 37
total market, addressable market vs., 205
touchpoints, brand, 73-74
Toyoda, Sakichi, 19
Toyota, 19
Track One, 280
traction, 206
trade groups, standard-setting by, 135
trade secrets, 275
trade shows, 254
trademarks, 274
traffic encryption, 108
transfer molding, 109
trending startups, 198
True Believers, 26-33
turning, 109

Tversky, Amos, 222
Twitter, 175, 236

U

UAVs (unmanned aerial vehicles), 14, 31, 143
UI (user interface), 88
UL (Underwriters Laboratories), 134
unaided sales, 249, 259
UnderArmour, 12
underwater robots, 14
Underwriters Laboratories (UL), 134
Unimate, 13
unintentional radiators, 290
unmanned aerial vehicles (UAVs), 14, 31, 143
updates, crowdfunding, 186
UPS, 266
user authentication, 108
user behaviors, 44
user characteristics, 44
user experience (UX), 83, 88
user interaction, 83
user interface (UI), 88
USPS, 266

V

value-based pricing, 226-228
value-based segmentation, 233
venture capital/capitalists, 201-209
　due diligence process, 208
　personalized introductions, 202-204
　targeting investors, 201
　telling your story to, 204-208
Venture Deals: Be Smarter Than Your Lawyer and Venture Capitalist (Feld and Mendelson), 210
venture debt, 212
vertical-focused incubators, 140
VholdR, 70
videos, crowdfunding, 173, 175
viral loops, 244
Virgin Group, 60
visual assets, 72
volume, sales, 225

Volusion, 262
VSC, 240
Vu, Sonny, 164-166, 171, 182, 228

W

warehousing and fulfillment, 261-266
Wareness, 182, 240
water jetting, 110
wave solder, 111
wearables (personal sensor devices), 11
website, 103
Wells Fargo, 260
WeMo, 10
Wheeler, Alina, 54, 61
Whiplash Logistics, 265
white-box packaging, 137
Whole Earth Catalog, 2
WiFi, 100
Wired (magazine), 11
wireframes, 88
Wolf, Gary, 11
working capital
　crowdfunding for, 187
　for manufacturing, 113
works-like prototypes, 85
Wozniak, Steve, 2
WSGR, 274

Y

Y Combinator (YC), 139, 177
year-over-year (YOY) trend, 38
yield rate, 133
Yocto, 96
Yoskovitz, Benjamin, 51
YouTube, 175

Z

Z-Wave, 99, 101
Zahn Center, 150
Zappos, 60
Zelman, Helen, 142
Zero to Maker (Lang), 26
ZigBee, 99, 101

About the Authors

Renee DiResta is Vice President of Business Development at Haven, a marketplace for ocean freight shipping, and a cofounder of the IoT Syndicate on AngelList. She was previously a principal at O'Reilly AlphaTech Ventures (OATV), where she spent four years as a VC investing in seed-stage technology startups. Prior to OATV, Renee spent seven years as an equity derivatives trader at Jane Street Capital, a quantitative proprietary trading firm in New York City. For fun, she plays with data sets, helps run The Maker Map open source project, and is an avid crafter. Renee holds a B.S. in computer science and political science from the Honors College of SUNY Stony Brook. She lives on the Web at *http://noupsi.de* and can be found on Twitter at @noupside.

Brady Forrest (@brady) runs Highway1, PCH's accelerator. You can see his day-to-day work with hardware startups on Syfy's Bazillion Dollar Club (to be released Fall 2015). He cofounded and shepherds Ignite, a global talk series, and is part of the team currently organizing Ignite SF. He's a venture advisor to 500 Startups and helps arts organizations via CAST-sf.org. Formerly, he worked on a number of things at O'Reilly Media, including the Radar blog, Web 2.0 Expo, Where 2.0, ETech, and Foo Camp. Most years, you can find him on the playa.

Ryan Vinyard is the Engineering Lead at Highway1, a hardware-focused startup accelerator located in San Francisco under parent company PCH International. He is a mechanical engineer who came to PCH through its consulting arm Lime Lab, where he developed consumer products for Fortune 500 brands. Previously, Ryan worked at startups in the cleantech and electric vehicle space, where he developed novel powertrain, motor control, and thermal systems. Ryan holds a B.S. in product design from Stanford University.

Colophon

The main title text was designed by Edie Freedman. Other cover fonts are URW Typewriter and Guardian Sans. The text font is Scala Pro, and the heading fonts are URW Typewriter and Benton Sans.

Have it your way.

O'Reilly eBooks

- Lifetime access to the book when you buy through oreilly.com
- Provided in up to four, DRM-free file formats, for use on the devices of your choice: PDF, .epub, Kindle-compatible .mobi, and Android .apk
- Fully searchable, with copy-and-paste, and print functionality
- We also alert you when we've updated the files with corrections and additions.

oreilly.com/ebooks/

Safari Books Online

- Access the contents and quickly search over 7000 books on technology, business, and certification guides
- Learn from expert video tutorials, and explore thousands of hours of video on technology and design topics
- Download whole books or chapters in PDF format, at no extra cost, to print or read on the go
- Early access to books as they're being written
- Interact directly with authors of upcoming books
- Save up to 35% on O'Reilly print books

See the complete Safari Library at safaribooksonline.com

O'REILLY®

Get even more for your money.

Join the O'Reilly Community, and register the O'Reilly books you own. It's free, and you'll get:

- $4.99 ebook upgrade offer
- 40% upgrade offer on O'Reilly print books
- Membership discounts on books and events
- Free lifetime updates to ebooks and videos
- Multiple ebook formats, DRM FREE
- Participation in the O'Reilly community
- Newsletters
- Account management
- 100% Satisfaction Guarantee

Signing up is easy:

1. Go to: oreilly.com/go/register
2. Create an O'Reilly login.
3. Provide your address.
4. Register your books.

Note: English-language books only

To order books online:
oreilly.com/store

For questions about products or an order:
orders@oreilly.com

To sign up to get topic-specific email announcements and/or news about upcoming books, conferences, special offers, and new technologies:
elists@oreilly.com

For technical questions about book content:
booktech@oreilly.com

To submit new book proposals to our editors:
proposals@oreilly.com

O'Reilly books are available in multiple DRM-free ebook formats. For more information:
oreilly.com/ebooks

©2014 O'Reilly Media, Inc. O'Reilly logo is a registered trademark of O'Reilly Media, Inc. 14373

CPSIA information can be obtained at www.ICGtesting.com
Printed in the USA
BVOW08s0314080615

403481BV00004B/5/P